The Open University

Mathematics/Science/Technology
An Inter-faculty Second Level Course
MST204 Mathematical Models and Methods

# Unit 15
# Newtonian mechanics
# in three dimensions

Prepared for the Course Team
by John Bolton

The Open University Press

The Open University Press, Walton Hall, Milton Keynes.

First published 1982. Reprinted 1983.

ISBN 0 335 14044 0

Produced in Great Britain by
Speedlith Photo Litho Limited, 76 Great Bridgewater Street, Manchester M1 5JY.

This text forms part of the correspondence element of an Open University Second Level Course.

For general availability of supporting material referred to in this text, please write to: Open University Educational Enterprises, 12 Cofferidge Close, Stony Stratford, Milton Keynes MK11 1BY, Great Britain.

Further information on Open University courses may be obtained from The Admissions Office, The Open University, P.O. Box 48, Milton Keynes MK7 6AB.

1.2

# Contents

# Introduction: taking stock

This unit returns to the subject of Newtonian mechanics. Several weeks have passed since you last studied mechanics so I shall briefly remind you of the story so far. There have been two fundamental ideas: Newton's second law, and the law of conservation of mechanical energy.

*Newton's second law* states that the total force is equal to mass times acceleration. We used it to predict the motion of a particle along a straight line. For example, we predicted the acceleration of a glider on a horizontal air-track (Section 2 of *Unit 4*) and the vertical oscillations of a bag of coins suspended by an elastic string (Section 1 of *Unit 7*).

*The law of conservation of mechanical energy* states that (under certain conditions) the sum of the kinetic and potential energies of a particle does not change with time. This was established, in *Unit 4*, by applying Newton's second law to a particle that moves along a straight line and experiences a total force that depends only on position. Later, in *Unit 7*, we assumed that the law of conservation of mechanical energy also applies to the motion of a particle along a curve, provided that influences like friction and air resistance can be ignored. This allowed us to explain oscillations of objects like a pendulum bob on a circular arc and a toy car on a cycloidal track (Section 3 of *Unit 7*).

These two laws have served us well. But it is important to realize that, so far, they have only been applied to a narrow range of phenomena. There have been three main limitations.

(1)   Newton's second law has only been applied to motion in a straight line. Yet straight-line motion is clearly the exception rather than the rule.

(2)   The law of conservation of mechanical energy has been applied in a rather ad hoc way: we gave no justification for extending it to motion in a curve and it was not entirely clear when this could be done.

(3)   We have always treated objects as if they were particles. This is not always a good approximation. For example, if a spinning top is modelled as a single particle many details of its motion are left unexplained.

We now begin the task of extending our treatment of Newtonian mechanics to remove these limitations. In this unit I shall concentrate on the first stage of generalization, and discuss Newton's second law for a particle moving along an arbitrary curve in three dimensions. I shall also give some justification for using the law of conservation of mechanical energy in systems like those in Section 3 of *Unit 7*. However, I shall say nothing about objects that cannot be modelled by particles, as this is the subject of *Units 17* and *29*.

## Study guide

This unit consists of *six* sections which should be studied in the order that they appear. I have split the subject matter into six, rather than five, sections in order to show the structure of the unit more clearly; however, some sections are shorter than usual, so that the unit should take about the same time to study as others in the course.

Section 1 covers many of the skills that are needed for Newtonian mechanics in more than one dimension. Some of this material will be revision for you, but I have collected it together in one place to make sure that you have it at your fingertips.

Section 2 contains the central ideas of the unit. Here, Newton's second law is introduced in vector form and the mechanics problems to be tackled in later sections are split into three classes.

Sections 3, 4 and 5 illustrate the three classes of problem with specific examples and exercises. Section 3 discusses particles that remain permanently at rest, Section 4 considers the tactics of a shot-putter who wishes to throw the shot as far as possible, and Section 5 discusses particles that move in circles, or arcs of circles.

The sixth section is a collection of problems for revision and further practice. An audio cassette tape forms an important part of Section 1 and the television programme provides the background to the problems of Section 4.

In studying this unit, you should make it your first priority to understand Sections 1 and 2. If you find these sections straightforward, they may take about 35% of your available time; otherwise, they may take up to 55%. The following table outlines study plans corresponding to each of these cases.

| Section | % time (plan 1) | % time (plan 2) |
|---------|-----------------|-----------------|
| 1 | 20 | 30 |
| 2 | 15 | 25 |
| 3 | 15 | 15 |
| 4 | 20 | 15 |
| 5 | 20 | 15 |
| 6 | 10 | 0 |

Plan 2 is based on the assumption that you treat some of the exercises in Sections 3–5 as worked examples, but it is important to spend some time on *each* of these sections, or your view of mechanics will be distorted.

# 1 The ingredients of Newton's second law in three dimensions (Tape Section)

In previous mechanics units you have used Newton's second law to predict the motion of particles along straight lines.

In one dimension, Newton's second law states that

$$m\ddot{x} = m\dot{v} = ma = F \tag{1}$$

where $m$ is the particle's mass (a positive constant),
    $x$ is the particle's position,
    $v$ is the velocity,
    $a$ is the acceleration,
    $F$ is the total force,

and each dot denotes a differentiation with respect to time.

One of the major aims of this unit is to extend Newton's second law to three dimensions so that you can use it to predict the motion of particles that do not move in straight lines. I hope it will not spoil the story for you if I reveal the conclusion we shall reach.

In three dimensions, Newton's second law states that

$$m\ddot{\mathbf{r}} = m\dot{\mathbf{v}} = m\mathbf{a} = \mathbf{F} \tag{2}$$

where $m$ is the particle's mass (a positive constant),
    $\mathbf{r}$ is the particle's position vector,
    $\mathbf{v}$ is the velocity vector,
    $\mathbf{a}$ is the acceleration vector,
    $\mathbf{F}$ is the total force vector,

and each dot denotes a differentiation with respect to time. This is remarkably similar to Equation (1): the only difference is that, in three dimensions, *vectors* are used to represent position, velocity, acceleration and force. That is why I shall spend much of this section reminding you about the position and velocity vectors and introducing you to two new vectors — the acceleration and force vectors.

## 1.1  A Cartesian co-ordinate system in mechanics

In order to quantify vectors I shall use a Cartesian co-ordinate system such as that shown in Figure 1. This consists of the following ingredients:

(1)   an origin, $O$;

(2)   three axes, the $x$-, $y$- and $z$-axes, which meet at the origin and point in mutually perpendicular directions;

(3)   three unit vectors, $\mathbf{i}$, $\mathbf{j}$ and $\mathbf{k}$, which point along the axes and which have unit magnitude.

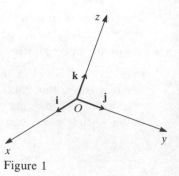

Figure 1

In setting up a Cartesian co-ordinate system we have a great deal of freedom. For example, the origin can be placed at any point and the $z$-axis can be aligned in any direction. However, in mechanics, it is conventional to restrict the choice of Cartesian co-ordinate system to those that have the following properties:

(1)   their axes are calibrated in SI units;

(2)   their axes are labelled in a right-handed way, with the $z$-axis being obtained from the $x$- and $y$-axes by the right-hand-screw rule;

(3)   they can be regarded as being static, with a fixed origin and with axes that maintain fixed directions.

The last of these points deserves special emphasis. You may recall that, in one-dimensional mechanics, the origin was always located at a fixed point. The reason was made clear in *Unit 8*: Newton's laws will not work if the origin itself is allowed to accelerate. In three-dimensional mechanics we must adopt a similar attitude and for a similar reason: Newton's laws will not work if the origin is allowed to accelerate or if the axes are allowed to rotate. That is why, in this unit, I shall always choose a co-ordinate system that is static relative to the earth.

You can think of the co-ordinate system as being like a signpost, with its origin at a definite place on the earth's surface and its axes picking out directions, like southwards, eastwards and upwards, that have a fixed geographical meaning. (Of course, it must be admitted that the earth is itself moving in a complicated way. Each year it completes one orbit around the sun, and each day it completes one revolution on its axis. So, strictly speaking, a signpost on earth steadily changes its orientation in the Universe as a whole. But this happens so slowly that it's almost imperceptible. I shall therefore use co-ordinate systems that are static on earth and make the approximation of neglecting their rotation and acceleration.)

*Unit 30* will have more to say about co-ordinate systems. The course *S354: Understanding Space and Time* also discusses these questions in great detail.

## 1.2  The position vector

Figure 2 reminds you how a Cartesian co-ordinate system is used to quantify the position of a particle. A displacement vector is drawn from the origin to the particle: this is known as the position vector and, in this unit, is denoted by the symbol $\mathbf{r}$.

The position vector can be expressed as a sum of scalar multiples of unit vectors:

$$\mathbf{r} = x\mathbf{i} + y\mathbf{j} + z\mathbf{k} \tag{3}$$

where the coefficients, $x$, $y$ and $z$, are known as the **components of the position vector** or, alternatively, as the **co-ordinates of the particle**. They are found by dropping perpendiculars from the particle onto each of the axes and reading off the numbers on their scales.

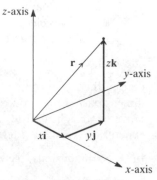

Figure 2

As time passes by the particle moves, so the position vector and its components are functions of time:

$$\mathbf{r}(t) = x(t)\mathbf{i} + y(t)\mathbf{j} + z(t)\mathbf{k}. \tag{4}$$

This vector function of time plays the same role here as the position function, $x(t)$, did in one dimension: it completely specifies the motion of the particle.

**Exercise 1**

The position vector of a particle is

$$\mathbf{r}(t) = \pi^2 t^2 \mathbf{i}$$

relative to a Cartesian co-ordinate system whose $x$-axis points along a horizontal air-track.

(i) Using the axes provided, mark the position of the particle at $t = 0$, $t = 0.25$, $t = 0.5$, $t = 0.75$ and $t = 1$.

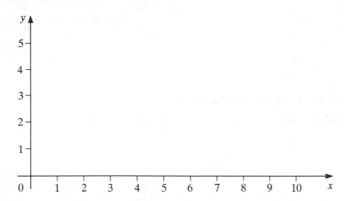

(ii) What can you say about the path and speed of the particle?

[*Solution on p. 45*]

**Exercise 2**

The position vector of a particle is

$$\mathbf{r}(t) = 2\cos\pi t\,\mathbf{i} + 2\sin\pi t\,\mathbf{j}$$

relative to a Cartesian co-ordinate system whose $z$-axis points vertically upwards.

(i) Using the axes provided, mark the position of the particle at $t = 0$, $t = 0.25$, $t = 0.5$, $t = 0.75$, $t = 1$, $t = 1.25$, $t = 1.5$ and $t = 1.75$.

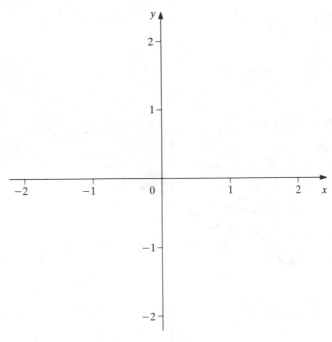

(ii) What can you say about the path and speed of the particle?

[*Solution on p. 45*]

## 1.3 The velocity and acceleration vectors

When you have completed Exercises 1 and 2 and checked with my solutions you should listen to the tape which explains how the position vector is used to find many other quantities of interest, including the velocity and acceleration vectors.

*Start the tape when you are ready.*

|  **Concepts** |  **Formulae** |
|---|---|
| 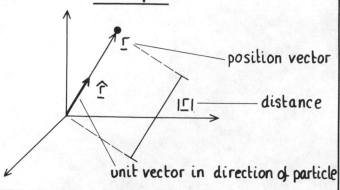 position vector | $\underline{r} = x\,\underline{i} + y\,\underline{j} + 3\,\underline{k}$ |
| $\lvert\underline{r}\rvert$ — distance | $\lvert\underline{r}\rvert = \sqrt{x^2 + y^2 + 3^2}$ |
| unit vector in direction of particle | $\hat{\underline{r}} = \dfrac{r}{\lvert\underline{r}\rvert}$ |

**1b** / **2b**

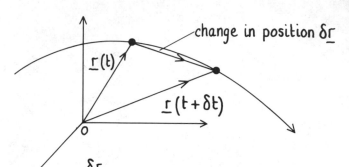

$$\frac{\delta\underline{r}}{\delta t} = \text{average velocity vector}$$

$$\lim_{\delta t \to 0} \frac{\delta\underline{r}}{\delta t} = \text{(instantaneous) velocity vector}$$

magnitude of velocity vector = speed

unit vector in direction of velocity

$$\delta\underline{r} = \delta x\,\underline{i} + \delta y\,\underline{j} + \delta 3\,\underline{k}$$

$$\underline{v} = \frac{\delta x\,\underline{i} + \delta y\,\underline{j} + \delta 3\,\underline{k}}{\delta t}$$

$$\underline{v} = \dot{\underline{r}} = \dot{x}\,\underline{i} + \dot{y}\,\underline{j} + \dot{3}\,\underline{k}$$

$$\lvert\underline{v}\rvert = \sqrt{\dot{x}^2 + \dot{y}^2 + \dot{3}^2}$$

$$\hat{\underline{v}} = \frac{\underline{v}}{\lvert\underline{v}\rvert}$$

**1c** / **2c**

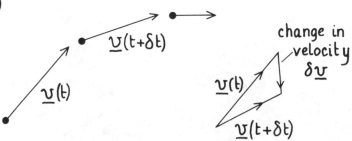

$$\lim_{\delta t \to 0} \frac{\delta\underline{v}}{\delta t} = \text{acceleration vector}$$

magnitude of acceleration vector

unit vector in direction of acceleration

$$\delta\underline{v} = \delta\dot{x}\,\underline{i} + \delta\dot{y}\,\underline{j} + \delta\dot{3}\,\underline{k}$$

$$\underline{a} = \ddot{\underline{r}} = \ddot{x}\,\underline{i} + \ddot{y}\,\underline{j} + \ddot{3}\,\underline{k}$$

$$\lvert\underline{a}\rvert = \sqrt{\ddot{x}^2 + \ddot{y}^2 + \ddot{3}^2}$$

$$\hat{\underline{a}} = \frac{a}{\lvert\underline{a}\rvert}$$

**3a** <u>Example 1</u>

$\underline{r} = \pi^2 t^2 \underline{i}$ $\qquad t > 0$

$|\underline{r}| = $ ☐

$\hat{\underline{r}} = $ ☐

**4a** <u>Example 2</u>

$\underline{r} = 2 \cos \pi t \, \underline{i} + 2 \sin \pi t \, \underline{j}$

$|\underline{r}| = $ ☐

$\hat{\underline{r}} = $ ☐

**3b**

$\delta \underline{r} = $ ☐

calculate between $t = 0$ and $t = 1$

$\overline{\underline{v}} = $ ☐

$\underline{v} = $ ☐

$|\underline{v}| = $ ☐

$\hat{\underline{v}} = $ ☐

**4b**

$\delta \underline{r} = $ ☐

calculate between $t = 0$ and $t = 1$

$\overline{\underline{v}} = $ ☐

$\underline{v} = $ ☐

$|\underline{v}| = $ ☐

$\hat{\underline{v}} = $ ☐

**3c**

$\underline{a} = $ ☐

$|\underline{a}| = $ ☐

$\hat{\underline{a}} = $ ☐

**4c**

$\underline{a} = $ ☐

$|\underline{a}| = $ ☐

$\hat{\underline{a}} = $ ☐

The tape raised two important points which I shall follow up here.

### (i)   Obtaining the velocity and acceleration vectors

I hope the tape has given you confidence in finding the velocity and acceleration vectors of a particle. To find the velocity vector you just differentiate each component of the position vector in turn, leaving the unit vectors $\mathbf{i}$, $\mathbf{j}$ and $\mathbf{k}$ unchanged:

$$\mathbf{v} = \dot{\mathbf{r}} = \dot{x}\mathbf{i} + \dot{y}\mathbf{j} + \dot{z}\mathbf{k}. \tag{5}$$

To find the acceleration vector you differentiate again:

$$\mathbf{a} = \ddot{\mathbf{r}} = \ddot{x}\mathbf{i} + \ddot{y}\mathbf{j} + \ddot{z}\mathbf{k}. \tag{6}$$

From this point of view, motion in three dimensions is almost as easy to describe as motion in one dimension. However, the dot notation introduces two ambiguities which I should clarify.

First, it is a nuisance that $\mathbf{i}$'s and $\mathbf{j}$'s are dotted as part of ordinary punctuation so that $\mathbf{i}$ means the unit vector along the $x$-axis, rather than its derivative with respect to time.

A more important ambiguity arises when we use the notation of dots for differentiation with respect to time and then write symbols like $|\dot{\mathbf{r}}|$ or $|\ddot{\mathbf{r}}|$. Does $|\dot{\mathbf{r}}|$ mean $\left|\dfrac{d\mathbf{r}}{dt}\right|$ or $\dfrac{d}{dt}|\mathbf{r}|$? In this course we shall adopt the convention that

$$|\dot{\mathbf{r}}| = \left|\frac{d\mathbf{r}}{dt}\right| \qquad \text{and} \qquad |\ddot{\mathbf{r}}| = \left|\frac{d^2\mathbf{r}}{dt^2}\right|$$

so that the dots always differentiate the vector *before* the magnitude is taken. It is important to stick to this convention because $\left|\dfrac{d\mathbf{r}}{dt}\right|$, $\dfrac{d}{dt}|\mathbf{r}|$, $\left|\dfrac{d^2\mathbf{r}}{dt^2}\right|$, $\dfrac{d}{dt}\left|\dfrac{d\mathbf{r}}{dt}\right|$ and $\dfrac{d^2}{dt^2}|\mathbf{r}|$ all represent *different* quantities:

$\left|\dfrac{d\mathbf{r}}{dt}\right| = |\dot{\mathbf{r}}|$ is the particle's speed;

$\dfrac{d}{dt}|\mathbf{r}|$ is the rate of change of the particle's distance from the origin;

$\left|\dfrac{d^2\mathbf{r}}{dt^2}\right| = |\ddot{\mathbf{r}}|$ is the magnitude of the particle's acceleration;

$\dfrac{d}{dt}\left|\dfrac{d\mathbf{r}}{dt}\right| = \dfrac{d}{dt}|\dot{\mathbf{r}}|$ is the rate of change of the particle's speed;

$\dfrac{d^2}{dt^2}|\mathbf{r}|$ is the rate of change of the rate of change of the particle's distance from the origin.

### Exercise 3
Use one of the examples discussed on the tape to show that, in general,

$$|\dot{\mathbf{r}}| \neq \frac{d}{dt}|\mathbf{r}| \qquad \text{and} \qquad |\ddot{\mathbf{r}}| \neq \frac{d}{dt}|\dot{\mathbf{r}}|.$$

[*Solution on p. 45*]

### (ii)   The significance of the acceleration vector

The second point arising from the tape concerns the significance of the acceleration vector. In everyday language 'acceleration' means increase in speed, but the acceleration vector is more subtle and more general than this for it describes both changes in speed *and* changes in the direction of motion.

The tape considered two extreme cases: motion at varying speed in a constant direction, and motion at constant speed in a varying direction, and in both cases the acceleration vector was non-zero. Let me emphasize this point again by asking you to look at the sketches in the solutions to Exercises 1 and 2. Here you see a striking image of what is being said — both these motions have the same constant magnitude of acceleration.

**Exercise 4**

Is there any moment between $t = 0$ and $t = 2$ when the particles in Exercises 1 and 2 have the same acceleration vector?

[*Solution on p. 45*]

## 1.4 Relationships between position, velocity and acceleration vectors

The purpose of this subsection is to give you practice at relating the concepts I have just introduced. If you already know the position vector and wish to find the velocity or acceleration vectors, the best strategy is to differentiate each component separately.

**Example 1**

The position vector of a particle varies with time according to the equation

$$\mathbf{r}(t) = -(t^3 + 3t)\mathbf{i} + 4t^2\mathbf{j} + 2\mathbf{k}.$$

Find (i)   the particle's velocity and acceleration vectors at $t = 0$;

   (ii) the particle's velocity and acceleration vectors at $t = 1$;

   (iii) the particle's speed and its direction of motion at $t = 1$.

*Solution*

Differentiating each component of $\mathbf{r}(t)$ in turn,

$$\mathbf{v}(t) = -(3t^2 + 3)\mathbf{i} + 8t\mathbf{j}. \tag{7}$$

Differentiating each component of $\mathbf{v}(t)$ in turn,

$$\mathbf{a}(t) = -6t\mathbf{i} + 8\mathbf{j}. \tag{8}$$

(i)   Substituting $t = 0$ in Equations (7) and (8), I obtain

$$\mathbf{v}(0) = -3\mathbf{i} \quad \text{and} \quad \mathbf{a}(0) = 8\mathbf{j}.$$

(ii)   Substituting $t = 1$ in Equations (7) and (8), I obtain

$$\mathbf{v}(1) = -6\mathbf{i} + 8\mathbf{j} \quad \text{and} \quad \mathbf{a}(1) = -6\mathbf{i} + 8\mathbf{j}.$$

(In this case, it happens that the vector $-6\mathbf{i} + 8\mathbf{j}$ represents both $\mathbf{v}(1)$ and $\mathbf{a}(1)$. This should not cause confusion — remember, we have adopted the convention of expressing all quantities in SI units. The above equations mean that, at $t = 1$, the velocity components in the $x$- and $y$-directions are $-6$ metres per second and 8 metres per second, while the corresponding acceleration components are $-6$ metres per second per second and 8 metres per second per second.)

(iii)   At $t = 1$ the particle's speed is

$$|\mathbf{v}(1)| = \sqrt{(-6)^2 + 8^2 + 0^2} = 10 \quad (\text{in ms}^{-1})$$

and it is moving in the direction of the unit vector

$$\hat{\mathbf{v}}(1) = \frac{-6\mathbf{i} + 8\mathbf{j}}{10} = -0.6\mathbf{i} + 0.8\mathbf{j}.$$

**Exercise 5**

The position vector of a particle varies with time according to the equation

$$\mathbf{r}(t) = 10t\,\mathbf{i} + (2 + 10t - 4.9t^2)\mathbf{j}.$$

Find (i)   the particle's speed at $t = 0$;

   (ii) the particle's direction of motion at $t = 0$;

   (iii) the particle's acceleration vector at any time, $t$.

[*Solution on p. 45*]

A different sort of problem arises when you know the acceleration vector and you wish to find the velocity or position vectors. In this case you should again treat each component separately but, instead of differentiating, you should integrate and use the given initial conditions.

**Example 2**

A particle has acceleration $\mathbf{a}(t) = 2\mathbf{i} - 6t\mathbf{k}$. At $t = 0$, it has velocity $\mathbf{v}(0) = 3\mathbf{i}$ and position $\mathbf{r}(0) = 8\mathbf{j}$.

(i)   Where is the particle at $t = 1$?

(ii)  How far from the origin is the particle at $t = 1$?

*Solution*

Integrating each component of $\mathbf{a}(t)$ in turn, and remembering to include arbitrary constants $A$, $B$ and $C$, I find

$$\mathbf{v}(t) = \left( \int 2\,dt + A \right)\mathbf{i} + \left( \int 0\,dt + B \right)\mathbf{j} + \left( \int -6t\,dt + C \right)\mathbf{k}$$
$$= (2t + A)\mathbf{i} + B\mathbf{j} + (-3t^2 + C)\mathbf{k}.$$

Integrating again, and remembering to include arbitrary constants $D$, $G$ and $H$, I find

$$\mathbf{r}(t) = \left( \int (2t + A)\,dt + D \right)\mathbf{i} + \left( \int B\,dt + G \right)\mathbf{j} + \left( \int (-3t^2 + C)\,dt + H \right)\mathbf{k}$$
$$= (t^2 + At + D)\mathbf{i} + (Bt + G)\mathbf{j} + (-t^3 + Ct + H)\mathbf{k}.$$

In mechanics I avoid using $E$ and $F$ for arbitrary constants, as they could be confused with energy and force.

Using the given initial conditions, $\mathbf{v}(0) = 3\mathbf{i}$ and $\mathbf{r}(0) = 8\mathbf{j}$, I conclude that

$$3\mathbf{i} = A\mathbf{i} + B\mathbf{j} + C\mathbf{k}$$

and      $8\mathbf{j} = D\mathbf{i} + G\mathbf{j} + H\mathbf{k}$

so       $A = 3$, $B = 0$, $C = 0$, $D = 0$, $G = 8$ and $H = 0$.

Substituting these values back into the expression for $\mathbf{r}(t)$ gives

$$\mathbf{r}(t) = (t^2 + 3t)\mathbf{i} + 8\mathbf{j} - t^3\mathbf{k}.$$

(i)   At $t = 1$, $\mathbf{r}(1) = 4\mathbf{i} + 8\mathbf{j} - \mathbf{k}$;

(ii)  At $t = 1$, $|\mathbf{r}(1)| = \sqrt{4^2 + 8^2 + (-1)^2} = 9$;

so, after 1 second the particle is 9 metres away from the origin.

**Exercise 6**

A particle has acceleration $\mathbf{a}(t) = (2t - 1)\mathbf{i} + \mathbf{j} + 2t\mathbf{k}$. At $t = 0$, it has velocity $\mathbf{v}(0) = -2\mathbf{i} - 2\mathbf{j} - 4\mathbf{k}$ and position $\mathbf{r}(0) = \frac{1}{3}\mathbf{i} - \frac{2}{3}\mathbf{k}$. Does the particle ever come instantaneously to rest? If so, *when* and at *what distance* from the origin?

[*Solution on p. 45*]

## 1.5   Individual force vectors

According to Equation (2), which summarizes Newton's second law in three dimensions, the acceleration vector of a particle is given by

$$\mathbf{a} = \frac{1}{m}\mathbf{F}. \qquad (2')$$

In Section 2 I shall make an attempt to justify this equation, but first I must explain the meaning of the quantities on the right-hand side: $m$, the mass of the particle and $\mathbf{F}$, the total force vector.

The first of these can be dealt with very quickly. Mass is taken to be a positive constant which does not depend on how the particle moves. For this reason, it has exactly the same value in three dimensions as in one dimension.

However, we shall certainly have to modify our description of forces. In one dimension all the forces acted along the same straight line. We were able to represent each force by a positive or negative number, and all these numbers were added together to yield the total force. But in three dimensions the forces can act in different directions and to specify the magnitude and direction of each force we need the language of vectors.

A typical situation is shown in Figure 3. Here, a particle of mass $m$ is attached to two stretched perfect springs which are anchored at different points $A$ and $B$ on a ceiling. There are three different forces acting on the particle: the force of gravity and forces due to each of the springs. All these forces act in different directions. The force of gravity acts downwards and the forces due to the stretched springs act from the particle towards the points $A$ and $B$.

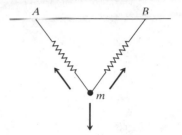

Figure 3

Before I can apply Newton's second law in such a situation, I must first do two things:

(i)    quantify the magnitude and direction of each force by a vector;

(ii)   combine the individual force vectors to obtain **F**, the total force vector, that describes the net effect of all the forces acting on the particle.

If you have studied *M101* you may know how to perform both these tasks. However, they are so important for your understanding of this unit that I shall spend some time on them now. In this subsection I shall consider the first task: that of finding individual force vectors. In the next subsection I shall combine these force vectors to obtain **F**.

I shall begin by discussing the force of gravity. You know from *Unit 4* that this force acts on all particles near the earth's surface: it acts downwards and has magnitude $mg$, where $m$ is the mass of the particle and $g = 9.8$ in SI units.

It is therefore very natural to represent the force of gravity by a vector which also points downwards and which also has magnitude $mg$. To be explicit I shall define the **gravitational force vector**, $\mathbf{F}_g$, as

$$\mathbf{F}_g = mg\,\hat{\mathbf{e}} \tag{9}$$

where $\hat{\mathbf{e}}$ is the unit vector that points downwards. However, Equation (9) is only the first step in quantifying the force of gravity. It is often necessary to choose a specific co-ordinate system and to express $\mathbf{F}_g$ in terms of the unit vectors **i**, **j** and **k** that point along the co-ordinate axes.

In some cases this can be done by inspection: for example, if the $y$-axis points vertically upwards, as in Figure 4,

$$\hat{\mathbf{e}} = -\mathbf{j}$$

so        $$\mathbf{F}_g = mg(-\mathbf{j}). \tag{10}$$

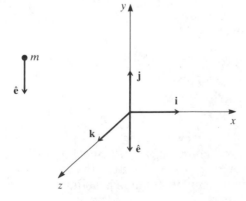

The two vectors labelled $\hat{\mathbf{e}}$ are equal because they have the same magnitudes and directions.

Figure 4

But we are not always so fortunate. The co-ordinate system may have been chosen (for other reasons) with no axis pointing vertically upwards or downwards; Figure 5 shows the general case. What is $\hat{\mathbf{e}}$ in terms of **i, j** and **k** here?

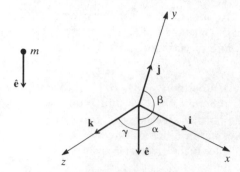

Figure 5

In order to answer this question I shall write

$$\hat{\mathbf{e}} = a\mathbf{i} + b\mathbf{j} + c\mathbf{k} \tag{11}$$

where $a$, $b$ and $c$ are numbers that are, as yet, unknown. They can be found by taking the scalar product of both sides of Equation (11) with $\mathbf{i}$, $\mathbf{j}$ and $\mathbf{k}$. Since $\mathbf{i}$, $\mathbf{j}$ and $\mathbf{k}$ are mutually perpendicular unit vectors, this gives

$$\mathbf{i}\cdot\hat{\mathbf{e}} = a, \qquad \mathbf{j}\cdot\hat{\mathbf{e}} = b, \qquad \mathbf{k}\cdot\hat{\mathbf{e}} = c.$$

Remembering the definition of the scalar product, I conclude that

$$a = \mathbf{i}\cdot\hat{\mathbf{e}} = \cos\alpha,$$
$$b = \mathbf{j}\cdot\hat{\mathbf{e}} = \cos\beta,$$
$$c = \mathbf{k}\cdot\hat{\mathbf{e}} = \cos\gamma,$$

where $\alpha$, $\beta$ and $\gamma$ are the angles shown in Figure 5. Substituting these results back into Equation (11) gives

$$\hat{\mathbf{e}} = \cos\alpha\,\mathbf{i} + \cos\beta\,\mathbf{j} + \cos\gamma\,\mathbf{k}. \tag{12}$$

Finally, Equations (12) and (9) allow us to write the gravitational force vector as

$$\mathbf{F}_g = mg(\cos\alpha\,\mathbf{i} + \cos\beta\,\mathbf{j} + \cos\gamma\,\mathbf{k}). \tag{13}$$

The angles $\alpha$, $\beta$ and $\gamma$ in Equations (12) and (13) are the angles between the directions of $\mathbf{i}$, $\mathbf{j}$ and $\mathbf{k}$ and the direction of $\hat{\mathbf{e}}$. I shall always measure these angles in radians and regard them as being positive so that

$$0 \leqslant \alpha \leqslant \pi, \qquad 0 \leqslant \beta \leqslant \pi \qquad \text{and} \qquad 0 \leqslant \gamma \leqslant \pi.$$

If these angles are known, it is easy to find $\hat{\mathbf{e}}$ by using the cosine function on your calculator.

## Example 3

Figure 6 shows a co-ordinate system in which the $z$-axis is horizontal, the $x$-axis points at an angle of $\dfrac{\pi}{6}$ radians below the horizontal, and the $y$-axis points at $\dfrac{\pi}{3}$ radians above the horizontal.

A particle of mass 3 kg experiences the force of gravity. Express this force as a vector in the co-ordinate system of Figure 6, giving your answer in newtons.

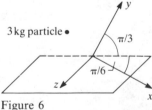

Figure 6

*Solution*

In Figure 7 I have represented the force of gravity by an arrow pointing downwards from the particle. However, it is difficult to find the angles $\alpha$, $\beta$ and $\gamma$ using this arrow, so I have drawn a second arrow pointing downwards from the origin. The second arrow has the same direction and magnitude as the first, so it represents the same vector. It is clear from Figure 7 that

$$\alpha = \frac{\pi}{2} - \frac{\pi}{6} = \frac{\pi}{3},$$

$$\beta = \frac{\pi}{2} + \frac{\pi}{3} = \frac{5\pi}{6},$$

$$\gamma = \frac{\pi}{2}.$$

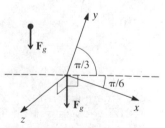

Figure 7

My calculator (set in the radian mode) then gives

$$\cos\alpha = \cos\frac{\pi}{3} = 0.5,$$

$$\cos\beta = \cos\frac{5\pi}{6} \simeq -0.866,$$

$$\cos\gamma = \cos\frac{\pi}{2} = 0,$$

so $\qquad \hat{\mathbf{e}} \simeq 0.5\mathbf{i} - 0.866\mathbf{j}.$

The magnitude of the gravitational force on the 3 kg particle is

$$|\mathbf{F}_g| = mg \simeq 3 \times 9.8 = 29.4,$$

Thus $\quad \mathbf{F}_g \simeq 29.4\,(0.5\mathbf{i} - 0.866\mathbf{j})$

$$\simeq 14.7\mathbf{i} - 25.5\mathbf{j} \qquad \text{(in newtons)}.$$

### Exercise 7

Repeat the calculation of Example 3, but this time find $\mathbf{F}_g$ in the co-ordinate system of Figure 8, where the z-axis is horizontal and the x-axis points at $\dfrac{\pi}{4}$ radians above the horizontal.

[Solution on p. 46]

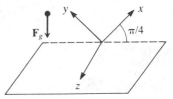

Figure 8

The methods we have just used are not restricted to the force of gravity. They illustrate a general procedure that can be used to quantify any force vector. The procedure consists of the following steps:

---

**Procedure 1.5: Representing a force by a vector**

1.  Draw a diagram showing the particle and your choice of Cartesian co-ordinate system.

2.  Draw an arrow from the particle in the direction of the force and label it with a symbol, e.g. $\mathbf{F}_1$.

3.  Specify the unit vector $\hat{\mathbf{F}}_1$ in the direction of the force. This is given by

    $$\hat{\mathbf{F}}_1 = \cos\alpha\,\mathbf{i} + \cos\beta\,\mathbf{j} + \cos\gamma\,\mathbf{k} \qquad (14)$$

    where $\alpha$, $\beta$ and $\gamma$ are the angles between the positive x-, y- and z-axes and your arrow.

    To find these angles it may be helpful to draw a second arrow in the same direction as the first, but with its tail at the origin.

4.  Specify the magnitude $|\mathbf{F}_1|$ of the force.

5.  Write down the **force vector** $\mathbf{F}_1$ by scaling the unit vector $\hat{\mathbf{F}}_1$ by the magnitude $|\mathbf{F}_1|$:

    $$\mathbf{F}_1 = |\mathbf{F}_1|\hat{\mathbf{F}}_1$$
    $$= |\mathbf{F}_1|(\cos\alpha\,\mathbf{i} + \cos\beta\,\mathbf{j} + \cos\gamma\,\mathbf{k}). \qquad (15)$$

---

When you do this yourself underline force vectors (instead of making them bold) to emphasize that they are vectors

Two comments should be made about this procedure. Firstly, there is a useful check that can be made after Step 3. The vector $\hat{\mathbf{F}}_1$ is of unit magnitude, so you should find that

$$\cos^2\alpha + \cos^2\beta + \cos^2\gamma = 1. \qquad (16)$$

If this is not so, you have either specified the wrong angles or taken their cosines incorrectly.

Secondly, it may happen that the angles $\alpha$, $\beta$ and $\gamma$ are not known numerically, but only in terms of some other variable. For example, Figure 9 shows two force vectors that act in the $x$, $y$-plane. (The $z$-axis has not been drawn as it is perpendicular to the page.)

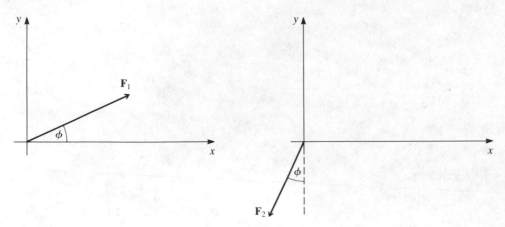

Figure 9

In this case we can still use the same procedure to conclude that

$$\mathbf{F}_1 = |\mathbf{F}_1|\left(\cos\phi\,\mathbf{i} + \cos\left(\frac{\pi}{2} - \phi\right)\mathbf{j}\right)$$

and      $$\mathbf{F}_2 = |\mathbf{F}_2|\left(\cos\left(\frac{\pi}{2} + \phi\right)\mathbf{i} + \cos(\pi - \phi)\mathbf{j}\right)$$

and these expressions can be simplified by using the trigonometric identities

$$\cos\left(\frac{\pi}{2} - \phi\right) = \sin\phi, \tag{17}$$

$$\cos\left(\frac{\pi}{2} + \phi\right) = -\sin\phi, \tag{18}$$

$$\cos(\pi - \phi) = -\cos\phi, \tag{19}$$

so that

$$\mathbf{F}_1 = |\mathbf{F}_1|(\cos\phi\,\mathbf{i} + \sin\phi\,\mathbf{j}), \tag{20}$$
$$\mathbf{F}_2 = |\mathbf{F}_2|(-\sin\phi\,\mathbf{i} - \cos\phi\,\mathbf{j}). \tag{21}$$

Procedure 1.5 is systematic and leads to the correct answers, but it is not the most direct way of obtaining Equations (20) and (21). It is quicker to realize that $\cos\alpha$, $\cos\beta$ and $\cos\gamma$ are the projections onto the $x$-, $y$- and $z$-axes of the unit vector in the direction of the force. Thus, Equations (20) and (21) follow immediately from the trigonometry of Figure 10.

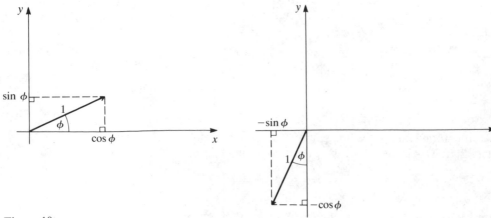

Figure 10

In the rest of this unit you may use either trigonometry *or* the method based on Equations (15) and (17)–(19). Once you are used to it, the trigonometric method is slightly quicker, but you may find that Equations (15) and (17)–(19) lead to fewer slips.

**Exercise 8**

Figure 11 shows two forces, $\mathbf{F}_3$ and $\mathbf{F}_4$, that act in the $x,y$-plane. Express these forces in terms of $\phi$ and the magnitudes $|\mathbf{F}_3|$ and $|\mathbf{F}_4|$.

[*Solution on p. 46*]

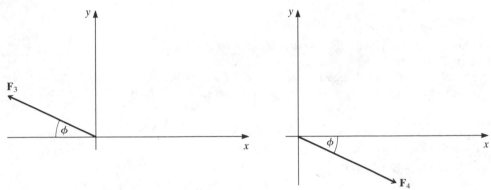

Figure 11

**Exercise 9**

Figure 12 shows the system introduced at the beginning of this subsection. The triangle $ABC$ lies in a vertical plane and each spring makes an angle of $\phi$ with the vertical direction and has length $l$. Each spring is perfect and has stiffness $k$ and natural length $l_0 < l$. Quantify the forces $\mathbf{F}_A$ and $\mathbf{F}_B$ on the particle $C$ due to each spring in the co-ordinate system of Figure 12. The springs lie in the $x,y$-plane and the $y$-axis points vertically upwards.

[*Solution on p. 46*]

By Hooke's law (*Unit 7*)
$$|\mathbf{F}_A| = |\mathbf{F}_B| = k(l - l_0)$$
and the force exerted at one end of a stretched perfect spring acts in a direction towards the other end of the spring.

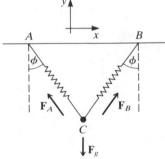

Figure 12

## 1.6  The total force vector

One question remains. What happens when several forces act at the same time on the same particle? For example, we have three force vectors, $\mathbf{F}_g$, $\mathbf{F}_A$ and $\mathbf{F}_B$ to represent the forces acting on the particle in Figure 12. But to use Newton's second law, I must specify a single force vector, $\mathbf{F}$, that represents the total force acting on the particle. So how should I combine the individual contributions $\mathbf{F}_g$, $\mathbf{F}_A$ and $\mathbf{F}_B$ to obtain the total force?

This is a deep question that can only be answered by performing experiments. Fortunately, these reveal a very simple pattern which is summarized by the following law:

---

**The law of addition of force**

Suppose a particle is acted on simultaneously by two or more forces that are represented by the vectors $\mathbf{F}_1, \mathbf{F}_2, \ldots$; then these forces have the same combined effect as a single force that is represented by the vector sum

$$\mathbf{F} = \mathbf{F}_1 + \mathbf{F}_2 + \cdots .  \tag{22}$$

The vector sum, $\mathbf{F}$, is known as the **total force vector** and it is this force vector that occurs in Newton's second law, $\mathbf{F} = m\mathbf{a}$.

---

The vector sum can be interpreted in terms of diagrams using the parallelogram rule. This is illustrated in Figure 13, where each arrow represents a force: the direction of the force is given by the direction of the arrow, and the magnitude of the force is proportional to the length of the arrow.

Figure 13(a) shows two forces $F_1$ and $F_2$ that act simultaneously on the same particle. Figure 13(b) shows the single force that produces the same net effect: it is found by drawing the diagonal of the parallelogram generated by $F_1$ and $F_2$.

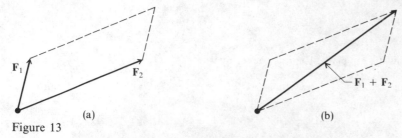

(a)                                                        (b)

Figure 13

For the special case of forces that act in the same direction, the parallelogram in Figure 13 'collapses' onto a straight line. However it is clear from Equation (22) that the magnitudes of these forces must be added together. Similarly, for forces that act in opposite directions along the same straight line, the magnitudes must be subtracted.

These special cases certainly agree with the rules given for combining forces in *Unit 4*, but Equation (22) also *extends* our previous work because it applies no matter what the directions of the forces and no matter how the particle moves.

**Exercise 10**

Figure 14 shows six particles that experience forces acting in the $x,y$-plane. It adopts the convention of representing each force by an arrow whose direction is that of the force and whose length is proportional to the magnitude of the force.

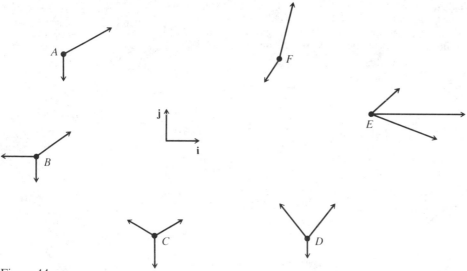

Figure 14

(i)     Which two particles experience zero total force?

(ii)    Which two particles experience a non-zero total force that acts in the direction of $i$?

(iii)   Which two particles experience a non-zero total force that acts in the direction of $j$?

[*Solution on p. 46*]

The parallelogram rule gives a clear intuitive picture of the addition of forces, but scale diagrams are often inaccurate and troublesome to construct in three dimensions, and a new parallelogram is needed whenever the forces change! Fortunately, we can avoid the need for diagrams by adding force vectors algebraically according to the rules of *Unit 14*: that is, we just add corresponding components. For example, the forces $F_1 = 2i + j + 3k$ and $F_2 = i + 3j - k$, acting simultaneously on the same particle, have the same effect as the force

$$F = F_1 + F_2$$
$$= (2i + j + 3k) + (i + 3j - k)$$
$$= 3i + 4j + 2k.$$

In general, you can use the following procedure for finding the total force acting on a particle.

---

**Procedure 1.6: Finding the total force acting on a particle**

1.  Draw a diagram showing the particle and your choice of Cartesian co-ordinate system.

2.  Draw an arrow in the direction of each force that acts on the particle. Label each arrow by a symbol, e.g. $\mathbf{F}_1, \mathbf{F}_2, \ldots$

3.  Quantify each force as in Procedure 1.5.

4.  Find the total force vector $\mathbf{F}$, by adding together the individual force vectors. This is done by adding together corresponding components. For example, the $x$-component of $\mathbf{F}$ is the sum of the $x$-components of $\mathbf{F}_1$, $\mathbf{F}_2, \ldots$ .

---

I shall write the components of the total force vector as $F_x$, $F_y$ and $F_z$ so that

$$\mathbf{F} = F_x \mathbf{i} + F_y \mathbf{j} + F_z \mathbf{k}.$$

Then, according to our procedures we have, for example,

$$F_x = \sum_i |\mathbf{F}_i| \cos \alpha_i$$

where $|\mathbf{F}_i|$ is the magnitude of the $i$th force, $\alpha_i$ is the angle between the direction of the $i$th force and the positive $x$-axis and the summation is over all the forces acting on the particle.

I use $F_x$, $F_y$, $F_z$ (rather than $F_1$, $F_2$, $F_3$ as in *Unit 14*) to avoid confusion with the individual forces $\mathbf{F}_1$, $\mathbf{F}_2$, $\mathbf{F}_3$.

**Exercise 11**

What is the total force experienced by the particle in Exercise 9? Write your answer as a vector expressed in the co-ordinate system of Figure 12.

[*Solution on p. 46*]

## Summary of Section 1

1.  In three-dimensional Newtonian mechanics the position of a particle is represented by the **position vector**

    $$\mathbf{r} = x\mathbf{i} + y\mathbf{j} + z\mathbf{k}$$

    where $x$, $y$ and $z$ are the **co-ordinates** of the particle and $\mathbf{i}$, $\mathbf{j}$, and $\mathbf{k}$ are unit vectors along the axes of a **static right-handed Cartesian co-ordinate system** that is calibrated in metres.

2.  The **velocity vector**, $\mathbf{v}$, is found by differentiating the position vector with respect to $t$, the time measured in seconds:

    $$\mathbf{v} = \dot{\mathbf{r}} = \dot{x}\mathbf{i} + \dot{y}\mathbf{j} + \dot{z}\mathbf{k}.$$

    The magnitude of the velocity vector is the **speed** of the particle. The direction of the velocity vector is the **direction of motion** of the particle.

3.  The **acceleration vector**, $\mathbf{a}$, is found by differentiating the velocity vector with respect to $t$:

    $$\mathbf{a} = \dot{\mathbf{v}} = \ddot{\mathbf{r}} = \ddot{x}\mathbf{i} + \ddot{y}\mathbf{j} + \ddot{z}\mathbf{k}.$$

    This describes the rate of change of velocity both in magnitude and direction. The acceleration will be non-zero provided *either* the speed *or* the direction of motion changes.

4.  Each individual force acting on a particle is represented by a **force vector**. If the direction of the force makes angles $\alpha$, $\beta$ and $\gamma$ with the positive $x$-, $y$- and $z$-axes and if the magnitude of the force is $|\mathbf{F}_1|$, then the force vector is

    $$\mathbf{F}_1 = |\mathbf{F}_1|(\cos\alpha\,\mathbf{i} + \cos\beta\,\mathbf{j} + \cos\gamma\,\mathbf{k}).$$

The cosines that appear in this equation can sometimes be simplified by using the trigonometric identities

$$\cos\left(\frac{\pi}{2} - \phi\right) = \sin\phi,$$

$$\cos\left(\frac{\pi}{2} + \phi\right) = -\sin\phi$$

and      $\cos(\pi - \phi) = -\cos\phi.$

5.  If a particle experiences a number of individual forces, represented by the vectors $\mathbf{F}_1, \mathbf{F}_2, \ldots$ it behaves just as if it were acted upon by the single force

$$\mathbf{F} = \mathbf{F}_1 + \mathbf{F}_2 + \cdots .$$

This formula is known as the **law of addition of forces** and the vector sum is known as the **total force vector**. The components of the total force vector are denoted by $F_x$, $F_y$ and $F_z$.

# 2   Newton's second law and three types of mechanics problem

## 2.1   A statement of Newton's second law

Section 1 assembled all the ingredients of Newtonian mechanics in more than one dimension:

the position vector, $\mathbf{r} = x\mathbf{i} + y\mathbf{j} + z\mathbf{k}$;

the velocity vector, $\dot{\mathbf{r}} = \dot{x}\mathbf{i} + \dot{y}\mathbf{j} + \dot{z}\mathbf{k}$;

the acceleration vector, $\ddot{\mathbf{r}} = \ddot{x}\mathbf{i} + \ddot{y}\mathbf{j} + \ddot{z}\mathbf{k}$;

the mass of a particle, $m$;

the total force vector, $\mathbf{F} = F_x\mathbf{i} + F_y\mathbf{j} + F_z\mathbf{k}$.

It is now time to give a precise statement of the physical law that binds these concepts together.

---

**Newton's second law: final formulation in terms of vectors**

Consider a particle of mass $m$. If, at a given instant of time the total force acting on the particle is represented by the vector $\mathbf{F}$, then the particle's acceleration vector at that instant is

$$\ddot{\mathbf{r}} = \frac{1}{m}\mathbf{F}.$$

That is,

$$m\ddot{\mathbf{r}} = \mathbf{F}. \tag{1}$$

---

In a chosen static Cartesian co-ordinate system Equation (1) can be written explicitly in terms of components:

$$m(\ddot{x}\mathbf{i} + \ddot{y}\mathbf{j} + \ddot{z}\mathbf{k}) = F_x\mathbf{i} + F_y\mathbf{j} + F_z\mathbf{k}.$$

Equating corresponding components on either side, we then see that Newton's second law is equivalent to the three scalar equations:

$$m\ddot{x} = F_x, \tag{2a}$$

$$m\ddot{y} = F_y, \tag{2b}$$

and      $m\ddot{z} = F_z, \tag{2c}$

## 2.2   Why Newton's second law?

Newton's second law, whether it is expressed as one vector equation or three
scalar equations, will form the basis of this unit and of all the remaining
mechanics units in this course. It is therefore natural to ask why this law should
be trusted. The most convincing reasons are provided by experiment. I shall make
no attempt to derive Equation (1) from more elementary ideas, but will regard it
as a fundamental hypothesis which must be judged by comparing its predictions
with reality.

Consider, for example, the motions discussed on the tape in Section 1. Suppose
particle 1, of mass $m$, is attached to a spring and pulled along a horizontal
frictionless track in such a way that its position vector is $\mathbf{r}(t) = \pi^2 t^2 \mathbf{i}$. Particle 2,
also of mass $m$, is tethered to a fixed point by an identical spring and swung in a
circle on a horizontal frictionless table in such a way that its position vector is
$\mathbf{r}(t) = 2\cos\pi t\,\mathbf{i} + 2\sin\pi t\,\mathbf{j}$.

Note that the spring attached
to particle 1 points along the
x-axis while the spring
attached to particle 2 points
radially inwards from the
particle to the origin.

During the tape session you calculated the acceleration vectors of these particles
and found that their directions were, respectively, $\mathbf{i}$ and $-\hat{\mathbf{r}}$. So, in both cases, the
acceleration vector points along the spring, in just the direction that one would
expect the force to act! The magnitudes of the acceleration vectors were also
calculated and found to be equal. Since the particles have the same mass, it follows
from Equation (1) that they should experience the same magnitude of total force.
However, the forces are provided by identical springs, so this can only happen if
both springs are extended by the same amount. This prediction can also be
checked by direct experiment.

Of course, this is only one illustration, but we can also use the experience of
generations of scientists who have found that Equation (1) is obeyed to a high
degree of accuracy under a wide range of conditions. In recent years, Equation (1)
has even been used to guide space probes to distant planets with pinpoint
accuracy. In conclusion, you can use Equation (1) with a great deal of confidence;
at no stage in this course will you need to question its validity.

## 2.3   A classification of mechanics problems

The rest of this unit is concerned with using Equations (1) and (2) to answer
questions about the real world. In order to gain a broad view of the tasks that lie
ahead, it is worth spending some time considering the different situations that can
arise.

Figure 1 (next page) shows my attempt to classify the main ways in which
Newton's second law is used. This classification is not exhaustive—but it does
cover all the applications you will meet in this unit.

According to Figure 1, there are three main types of problem:

(1)   Problems in which the acceleration vector is known, and Newton's second
law allows us to predict a force by means of ordinary *algebraic* equations.

(2)   Problems in which the total force vector is known, and Newton's second law
allows us to predict the particle's motion by means of *differential* equations
that are *uncoupled* from one another.

(3)   Problems in which the total force vector is known, and Newton's second law
allows us to predict the particle's motion by means of *differential* equations
that are *coupled* together.

The easiest way of understanding this classification is to look at some specific
examples.

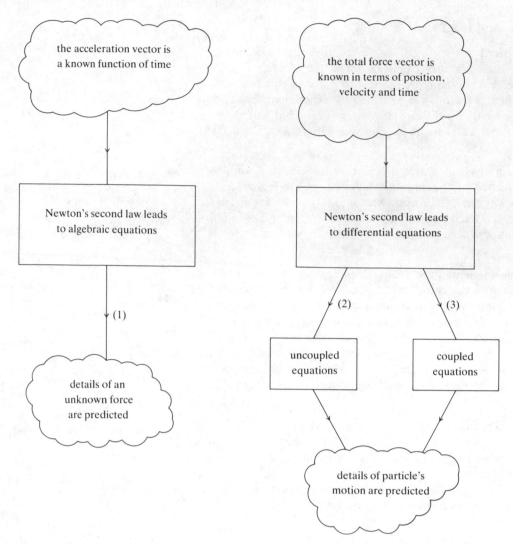

Figure 1

## 2.4   Some sample problems

### Example 1

A particle of mass $m$ has position vector

$$\mathbf{r}(t) = l\cos\omega t\,\mathbf{i} + l\sin\omega t\,\mathbf{j} \tag{3}$$

where $l$ and $\omega$ are positive constants. Find the magnitude and direction of the total force acting on the particle.

*Solution*

In this case, it is easy to find the acceleration vector of the particle by differentiating Equation (3) twice. This gives

$$\dot{\mathbf{r}}(t) = -l\omega\sin\omega t\,\mathbf{i} + l\omega\cos\omega t\,\mathbf{j};$$
$$\ddot{\mathbf{r}}(t) = -l\omega^2\cos\omega t\,\mathbf{i} - l\omega^2\sin\omega t\,\mathbf{j}. \tag{4}$$

Hence, using Newton's second law, the total force acting on the particle is

$$\mathbf{F} = m\ddot{\mathbf{r}}$$
$$= ml\omega^2(-\cos\omega t\,\mathbf{i} - \sin\omega t\,\mathbf{j}).$$

However, the unit vector in the direction of $\mathbf{r}$ is

$$\hat{\mathbf{r}} = \frac{\mathbf{r}}{|\mathbf{r}|} = \cos\omega t\,\mathbf{i} + \sin\omega t\,\mathbf{j}$$

so,    $$\mathbf{F} = ml\omega^2(-\hat{\mathbf{r}}). \tag{5}$$

That is, the total force has magnitude $ml\omega^2$ and acts in the direction of $-\hat{\mathbf{r}}$, from the particle towards the origin.

**Exercise 1**

(i)   Show that the particle described in Example 1 moves at constant speed $l\omega$ in a circle of radius $l$, and takes a time $T = 2\pi/\omega$ to complete one orbit.

(ii)  Suppose that the force calculated in Example 1 is provided by a perfect spring of stiffness $k$ which is anchored at the centre of the circle. Show that the time $T$ must be greater than $2\pi\sqrt{m/k}$ no matter how much the spring is stretched.

Because its speed is constant the particle is said to perform **uniform circular motion**.

[Solution on p. 46]

The important thing to notice about Example 1 is that Newton's second law was used as an *algebraic equation* to find an unknown force from a known acceleration. The result (Equation (5)) was then used to solve Exercise 1(ii). This illustrates the first type of mechanics problem, represented by the left-hand arrow in Figure 1.

We now turn to consider the second type of problem, involving *uncoupled differential* equations.

**Example 2**

A particle of mass $m$ is acted on by two forces. The force of gravity, $\mathbf{F}_g$, acts downwards. A second force, $\mathbf{F}_e$, acts from the particle towards a fixed point, $O$, and has a magnitude equal to the particle's distance from $O$, multiplied by a positive constant, $c$. If the particle starts at $t = 0$, with position $\mathbf{r}(0)$ and velocity $\dot{\mathbf{r}}(0)$, what is its position at any subsequent time, $t$?

*Solution*

In this case, it is easy to find the total force acting on the particle by using Procedure 1.6.

My diagram is shown in Figure 2. Notice that I have chosen a co-ordinate system in which one axis (the $x$-axis) points vertically downwards: this will simplify the description of $\mathbf{F}_g$. I have also chosen the origin of my co-ordinate system at the fixed point, $O$: this will simplify the description of $\mathbf{F}_e$.

With this choice of co-ordinate system, the directions of the forces are found, by inspection, to be

$$\hat{\mathbf{F}}_g = \mathbf{i},$$
$$\hat{\mathbf{F}}_e = -\hat{\mathbf{r}}$$
$$= -\frac{\mathbf{r}}{|\mathbf{r}|} = -\frac{(x\mathbf{i} + y\mathbf{j} + z\mathbf{k})}{|\mathbf{r}|}.$$

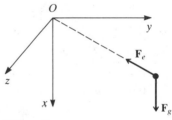

Figure 2

The magnitudes of the forces are

$$|\mathbf{F}_g| = mg,$$
$$|\mathbf{F}_e| = c|\mathbf{r}|.$$

Thus

$$\mathbf{F}_g = mg\,\mathbf{i}$$

and    $$\mathbf{F}_e = -c|\mathbf{r}|\frac{(x\mathbf{i} + y\mathbf{j} + z\mathbf{k})}{|\mathbf{r}|} = -cx\mathbf{i} - cy\mathbf{j} - cz\mathbf{k}.$$

The total force acting on the particle is therefore

$$\mathbf{F} = \mathbf{F}_g + \mathbf{F}_e$$
$$= (mg - cx)\mathbf{i} - cy\mathbf{j} - cz\mathbf{k}.$$

Using this result in Newton's second law (Equations (2a)–(2c)) I obtain the three *differential* equations:

$$m\ddot{x} = mg - cx, \tag{6a}$$
$$m\ddot{y} = -cy, \tag{6b}$$
$$m\ddot{z} = -cz. \tag{6c}$$

On setting $\omega^2 = \dfrac{c}{m}$, and rearranging, I obtain

$$\ddot{x} + \omega^2 x = g, \tag{7a}$$
$$\ddot{y} + \omega^2 y = 0, \tag{7b}$$
$$\ddot{z} + \omega^2 z = 0. \tag{7c}$$

The important thing to notice about these equations is that they are **uncoupled** — that is, they can be solved independently of one another. Fortunately, they also have a very familiar form which you will recognize from *Unit 7* on simple harmonic oscillations. We can immediately write down their general solutions as

$$x(t) = C_1 \cos\omega t + D_1 \sin\omega t + \frac{g}{\omega^2},$$

$$y(t) = C_2 \cos\omega t + D_2 \sin\omega t,$$

$$z(t) = C_3 \cos\omega t + D_3 \sin\omega t$$

where $C_1$, $C_2$, $C_3$, $D_1$, $D_2$ and $D_3$ are arbitrary constants. The position vector of the particle is therefore

$$\begin{aligned}
\mathbf{r}(t) &= x(t)\mathbf{i} + y(t)\mathbf{j} + z(t)\mathbf{k} \\
&= \left( C_1 \cos\omega t + D_1 \sin\omega t + \frac{g}{\omega^2} \right)\mathbf{i} + (C_2 \cos\omega t + D_2 \sin\omega t)\mathbf{j} \\
&\quad + (C_3 \cos\omega t + D_3 \sin\omega t)\mathbf{k}
\end{aligned}$$

which can be rewritten as

$$\mathbf{r}(t) = \cos\omega t\, \mathbf{C} + \sin\omega t\, \mathbf{D} + \frac{g}{\omega^2}\mathbf{i} \tag{8}$$

where   $\mathbf{C} = C_1\mathbf{i} + C_2\mathbf{j} + C_3\mathbf{k}$

and   $\mathbf{D} = D_1\mathbf{i} + D_2\mathbf{j} + D_3\mathbf{k}$

are arbitrary constant vectors which must be chosen to agree with the given initial conditions. Substituting $t = 0$ in Equation (8) gives

$$\mathbf{r}(0) = \mathbf{C} + \frac{g}{\omega^2}\mathbf{i}$$

so   $\mathbf{C} = \mathbf{r}(0) - \dfrac{g}{\omega^2}\mathbf{i}.$ \hfill (9)

Differentiating Equation (8) and setting $t = 0$ gives

$$\dot{\mathbf{r}}(0) = \omega\mathbf{D}$$

so   $\mathbf{D} = \dfrac{1}{\omega}\dot{\mathbf{r}}(0).$ \hfill (10)

Finally, combining Equations (8), (9) and (10), I see that the position of the particle at any time $t \geqslant 0$ is

$$\mathbf{r}(t) = \cos\omega t\, \mathbf{r}(0) + \frac{1}{\omega}\sin\omega t\, \dot{\mathbf{r}}(0) + \frac{g}{\omega^2}(1 - \cos\omega t)\mathbf{i} \tag{11}$$

where $\omega = \sqrt{\dfrac{c}{m}}$.

The above example illustrates the power of our new vector methods. The next exercise will reveal the wealth of information contained in Equation (11): depending on the initial conditions, it tells us exactly which motion will occur.

### Exercise 2

Consider the particle in Example 2, and let $A$ and $B$ be constants.

(i)   Show that the particle remains permanently at rest if

$$\mathbf{r}(0) = \frac{g}{\omega^2}\mathbf{i} \qquad \text{and} \qquad \dot{\mathbf{r}}(0) = \mathbf{0},$$

and oscillates simple-harmonically along a vertical straight line if

$$\mathbf{r}(0) = \left(\frac{g}{\omega^2} + A\right)\mathbf{i} \qquad \text{and} \qquad \dot{\mathbf{r}}(0) = \omega B\mathbf{i}.$$

(ii)    Describe the motion of the particle for the following sets of initial conditions:

(a)   $\mathbf{r}(0) = \dfrac{g}{\omega^2}\mathbf{i} + A\mathbf{j} \qquad \text{and} \qquad \dot{\mathbf{r}}(0) = \omega B\mathbf{j}$;

(b)   $\mathbf{r}(0) = \dfrac{g}{\omega^2}\mathbf{i} + A\mathbf{j} \qquad \text{and} \qquad \dot{\mathbf{r}}(0) = \omega A\mathbf{k}$;

(c)   $\mathbf{r}(0) = \dfrac{g}{\omega^2}\mathbf{i} + A\mathbf{j} \qquad \text{and} \qquad \dot{\mathbf{r}}(0) = \omega A\mathbf{i}$.

[*Solution on p. 47*]

In Example 2 and Exercise 2, Newton's second law was used as a set of *uncoupled differential equations*, relating $\ddot{x}$ to $x$, $\ddot{y}$ to $y$ and $\ddot{z}$ to $z$. These differential equations were solved, subject to given initial conditions, and the position vector of the particle was predicted. This illustrates the second type of mechanics problem, represented by the middle arrow (2) in Figure 1.

In many ways, a problem of this type resembles *three* one-dimensional problems because Newton's second law gives rise to *three* equations which can be tackled one at a time. Unfortunately, many problems in mechanics are more troublesome than this:

## Example 3

Physicists inform us that any electrically charged particle, moving with velocity $\mathbf{v}$ in a constant magnetic field, experiences a force

$$\mathbf{F}_B = q(\mathbf{v} \times \mathbf{B})$$

where $q$ is a number representing the electric charge of the particle and $\mathbf{B}$ is a vector representing the magnetic field. Use this fact to predict the motion of a particle of charge $q = 1.6 \times 10^{-19}$ and mass $m = 1.6 \times 10^{-24}$, under the following circumstances:

$$\mathbf{B} = b\mathbf{k}, \qquad \text{where } b = 10^{-2};$$

$$\mathbf{r}(0) = 2\mathbf{j};$$

$$\dot{\mathbf{r}}(0) = 2000\mathbf{i}.$$

You may neglect all forces other than $\mathbf{F}_B$.

It is not necessary to understand electricity or magnetism to answer this question. However you may know that a television picture is produced when electrons strike the screen. Electrons are charged particles and the force $\mathbf{F}_B$ can be demonstrated by bringing a magnet close to the set.

*Solution*

We are told that the total force acting on the particle is

$$\mathbf{F} = \mathbf{F}_B = q(\mathbf{v} \times \mathbf{B}).$$

Evaluating the vector product, I have

$$\begin{aligned}
\mathbf{F} &= q(\dot{x}\mathbf{i} + \dot{y}\mathbf{j} + \dot{z}\mathbf{k}) \times (b\mathbf{k}) \\
&= qb(\dot{x}\mathbf{i} \times \mathbf{k} + \dot{y}\mathbf{j} \times \mathbf{k} + \dot{z}\mathbf{k} \times \mathbf{k}) \\
&= qb(-\dot{x}\mathbf{j} + \dot{y}\mathbf{i}).
\end{aligned}$$

So Newton's second law gives

$$m\ddot{\mathbf{r}} = qb(\dot{y}\mathbf{i} - \dot{x}\mathbf{j})$$

or      $\ddot{\mathbf{r}} = \ddot{x}\mathbf{i} + \ddot{y}\mathbf{j} + \ddot{z}\mathbf{k} = \Omega(\dot{y}\mathbf{i} - \dot{x}\mathbf{j})$

where $\Omega = \dfrac{qb}{m} = \dfrac{1.6 \times 10^{-19} \times 10^{-2}}{1.6 \times 10^{-24}} = 1000.$

Taking each component separately, I have

$$\ddot{x} = \Omega\dot{y}; \tag{12a}$$

$$\ddot{y} = -\Omega\dot{x}; \tag{12b}$$

$$\ddot{z} = 0. \tag{12c}$$

If you compare these equations with Equations (7a)–(7c) you will see that an extra complication has arisen. Equations (12a) and (12b) are **coupled** together. They each involve $x$ *and* $y$ so they cannot be tackled independently, but must first be unravelled. It is sometimes difficult to spot how to do this, but in the present case I simply need to differentiate Equation (12a) to give

$$\ddot{x} = \Omega\ddot{y}$$

Then, using Equation (12b), I find that

$$\dddot{x} = -\Omega^2\dot{x}$$

or    $$\frac{d^2}{dt^2}\dot{x} + \Omega^2\dot{x} = 0.$$

The left-hand side of this equation is $x$ *triple* dot, the *third* derivative of $x$ with respect to time.

This is a second-order differential equation for the $x$-component of the velocity. Its general solution is

$$\dot{x}(t) = C\cos\Omega t + D\sin\Omega t \tag{13}$$

where $C$ and $D$ are arbitrary constants.

The $y$-component of the velocity is found by combining Equations (12a) and (13) to obtain

$$\dot{y}(t) = \frac{1}{\Omega}\ddot{x}(t)$$

$$= \frac{1}{\Omega}\frac{d}{dt}(C\cos\Omega t + D\sin\Omega t)$$

$$= -C\sin\Omega t + D\cos\Omega t. \tag{14}$$

The $z$-component of the velocity is found by integrating Equation (12c) to obtain

$$\dot{z}(t) = H \tag{15}$$

where $H$ is an arbitrary constant.

Thus, combining Equations (13), (14) and (15), the velocity vector of the particle is

$$\dot{\mathbf{r}}(t) = (C\cos\Omega t + D\sin\Omega t)\mathbf{i} + (-C\sin\Omega t + D\cos\Omega t)\mathbf{j} + H\mathbf{k}.$$

Comparing this with the given initial condition $\dot{\mathbf{r}}(0) = 2000\,\mathbf{i}$, I conclude that $C = 2000$, $D = 0$ and $H = 0$.

Hence,

$$\dot{\mathbf{r}}(t) = 2000\cos(1000t)\,\mathbf{i} - 2000\sin(1000t)\,\mathbf{j}.$$

Integrating both sides of this equation gives

$$\mathbf{r}(t) = 2\sin(1000t)\mathbf{i} + 2\cos(1000t)\mathbf{j} + \mathbf{G}$$

where $\mathbf{G}$ is a constant vector.

Comparing this with the given initial condition $\mathbf{r}(0) = 2\mathbf{j}$, I conclude that $\mathbf{G} = \mathbf{0}$ so that

$$\mathbf{r}(t) = 2\sin(1000t)\,\mathbf{i} + 2\cos(1000t)\,\mathbf{j}.$$

Thus the particle moves at constant speed ($2000\,\text{ms}^{-1}$) in a circle of radius 2 metres.

This follows by an argument similar to that used in Exercise 1 (i).

**Exercise 3**

Consider the particle in Example 3, but this time include the effects of both the magnetic force, $q(\mathbf{v} \times \mathbf{B})$ *and* the force of gravity. Choose a co-ordinate system whose $y$-axis points vertically upwards and predict the motion of the particle for the same values of $q$, $m$, $\mathbf{B}$, $\mathbf{r}(0)$ and $\dot{\mathbf{r}}(0)$ as before.

[*Solution on p. 47*]

Example 3 and Exercise 3 illustrate the third type of mechanics problem, in which Newton's second law leads to differential equations that are coupled together. These problems are represented by the right-hand arrow in Figure 1. Notice that the level of complexity steadily increases as we pass from Example 1 to Example 2 to Example 3. These three levels of complexity dictate the structure of the rest of this unit.

Section 3 discusses problems in which Newton's second law leads to algebraic equations. In Section 4, Newton's second law leads to differential equations that are solved independently of one another, and in Section 5 the differential equations are coupled together. The classification given in Figure 1 is therefore a sort of route map that will guide you through the next three sections.

### Summary of Section 2

1.  **Newton's second law**, in its final vector formulation, states that, at each instant of time, the motion of a particle is governed by the equation

    $$m\ddot{\mathbf{r}} = \mathbf{F}$$

    or, in terms of components,

    $$m\ddot{x} = F_x,$$
    $$m\ddot{y} = F_y,$$
    and $\quad m\ddot{z} = F_z,$

    where $\quad m$ is the mass of the particle,

    $\ddot{\mathbf{r}} = \ddot{x}\mathbf{i} + \ddot{y}\mathbf{j} + \ddot{z}\mathbf{k}$ is the **acceleration vector**,

    and $\quad \mathbf{F} = F_x\mathbf{i} + F_y\mathbf{j} + F_z\mathbf{k}$ is the **total force vector**.

2.  The mechanics problems in this unit can be split into three main classes:

    (1)  Problems in which Newton's second law leads to **algebraic equations** (e.g. Example 1).

    (2)  Problems in which Newton's second law leads to **uncoupled differential equations** (e.g. Example 2)

    (3)  Problems in which Newton's second law leads to **coupled differential equations** (e.g. Example 3).

3.  A particle that moves in a circle with constant speed is said to perform **uniform circular motion**. For example, a particle of mass $m$ with position vector

    $$\mathbf{r}(t) = l\cos\omega t\,\mathbf{i} + l\sin\omega t\,\mathbf{j}$$

    performs uniform circular motion at speed $l\omega$ round a circle of radius $l$. The total force acting on this particle has magnitude $ml\omega^2$ and points from the particle towards the centre of the circle.

# 3    Algebraic equations and sloping tables

You have had a glimpse of the three main classes of mechanics problem. This section will take a closer look at the first class of problem, in which Newton's second law is treated as an algebraic equation, relating an unknown force to a known acceleration.

## 3.1    Static particles

The simplest case involves particles that remain **static** (i.e. permanently at rest). The position vector of a static particle does not vary with time so $\ddot{\mathbf{r}} = \mathbf{0}$ and, by Newton's second law,

$$\mathbf{F} = \mathbf{0}. \tag{1}$$

That is, for a static particle, the total force, obtained by adding together all the forces acting on the particle, is equal to the zero vector. The precise meaning of this statement is examined in the following exercise:

**Exercise 1**

Consider the six particles in Exercise 10 of Section 1.

(i)   Which of the particles *could* be static?
(ii)  Which of the particles *must* be static?
(iii) Which of the particles *could* be momentarily at rest?
(iv)  Which of the particles *must* be momentarily at rest?

[*Solution on p. 48*]

In order to apply Equation (1), we must first find the total force in terms of other variables, using Procedure 1.6.

**Example 1**

Figure 1 shows the system discussed in Exercises 9 and 11 of Section 1. The spring has stiffness 100 newtons per metre and natural length 1.0 metre, and the distance $AB$ is 2.2 metres. When the particle is static, the distances $AC$ and $BC$ are both 1.2 metres. What is the mass of the particle?

*Solution*

The total force acting on the particle has already been calculated in Exercise 11 of Section 1. It is

$$\mathbf{F} = (2k(l - l_0)\cos\phi - mg)\mathbf{j}.$$

Setting $\mathbf{F} = \mathbf{0}$ gives

$$m = \frac{2k}{g}(l - l_0)\cos\phi.$$

According to the data given in the question, $k = 100$ and $l - l_0 = 1.2 - 1.0 = 0.2$. Also

$$\cos\phi = \frac{DC}{AC} = \frac{\sqrt{(AC)^2 - (AD)^2}}{AC} = \frac{\sqrt{(1.2)^2 - (1.1)^2}}{1.2} \simeq 0.4$$

and $g = 9.8$ so

$$m \simeq \frac{2 \times 100 \times 0.2 \times 0.4}{9.8} \simeq 1.63.$$

The mass of the particle is 1.63 kilograms.

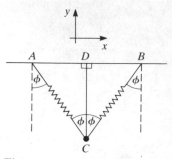

Figure 1

**Exercise 2**

Figure 2 shows a Cartesian co-ordinate system with a horizontal $x$-axis and a vertical $y$-axis. A particle of mass $m$ experiences three forces, each acting in the $x,y$-plane:

$\mathbf{F}_g$ is the force due to gravity. It acts downwards and has magnitude $mg$.

$\mathbf{F}_1$ is the force due to a taut string. It acts in the direction from the particle to the other end of the string, fixed at $A$, and has an unknown magnitude, $|\mathbf{F}_1|$. The angle between the string and the vertical is $\pi/4$ radians.

$\mathbf{F}_2$ is the force due to a stretched spring. It acts horizontally and has an unknown magnitude, $|\mathbf{F}_2|$.

(i)   Find the total force acting on the particle in terms of $mg$, $|\mathbf{F}_1|$ and $|\mathbf{F}_2|$.

(ii)  When the particle is static, what are the values of $|\mathbf{F}_1|$ and $|\mathbf{F}_2|$?

[*Solution on p. 48*]

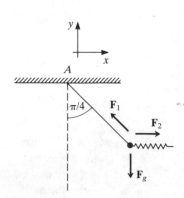

Figure 2

## 3.2  Contact forces

Some of the most common examples of static particles arise when an object is in contact with a solid surface. Consider, first, the situation illustrated in Figure 3, which shows an object lying at rest on a horizontal surface. The object is pulled downwards by the force of gravity, $\mathbf{F}_g$, so why doesn't it accelerate downwards? In *Unit 4* we merely stated that the horizontal surface prevents it from doing so. Now we can be more specific.

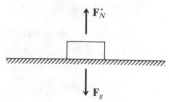

Figure 3

Equation (1) tells us that a static particle experiences zero total force. This means that the horizontal surface must push the object upwards with a force, $\mathbf{F}_N$, such that

$$\mathbf{F}_g + \mathbf{F}_N = \mathbf{0}.$$

The force $\mathbf{F}_N$ is known as the **normal reaction** because the surface *reacts* against being compressed and pushes the object upwards, in a direction *normal* (i.e. perpendicular) to the surface.

Even if the surface slopes or the particle moves, a normal reaction exists. For example, Figure 4 shows a glider that rides on a cushion of air just above a tilted table. I have indicated the forces on the glider: gravity acts downwards and the normal reaction acts perpendicular to the table. The magnitude of the normal reaction force is not known yet, *but it must be just sufficient to prevent the glider from sinking into the table.* I shall call this condition the **normal reaction rule**; the following example will show how it is used.

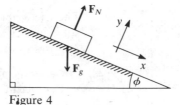

Figure 4

### Example 2

Find the magnitude of the normal reaction force in terms of the mass, $m$, of the glider and the angle $\phi$ in Figure 4.

*Solution*

In Figure 4, I have chosen a co-ordinate system with an $x$-axis that points directly down the slope and a $y$-axis that is perpendicular to the surface. The $z$-axis points towards you, out of the page. (This choice of co-ordinate system is convenient because the $y$- and $z$-co-ordinates of the glider remain constant.) In my co-ordinate system, both forces act in the $x,y$-plane and, from Figure 5, their directions are

$$\hat{\mathbf{F}}_g = \sin\phi\,\mathbf{i} - \cos\phi\,\mathbf{j} \qquad \text{and} \qquad \hat{\mathbf{F}}_N = \mathbf{j}.$$

Thus, the total force acting on the glider is

$$\begin{aligned}\mathbf{F} &= mg(\sin\phi\,\mathbf{i} - \cos\phi\,\mathbf{j}) + |\mathbf{F}_N|\mathbf{j} \\ &= mg\sin\phi\,\mathbf{i} + (|\mathbf{F}_N| - mg\cos\phi)\mathbf{j}. \end{aligned} \qquad (2)$$

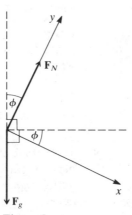

Figure 5

Using the $y$-component of Newton's second law, we have

$$m\ddot{y} = F_y = |\mathbf{F}_N| - mg\cos\phi.$$

However, we know that the glider does not sink into the table, so that $\ddot{y} = 0$. Hence,

$$|\mathbf{F}_N| - mg\cos\phi = 0$$

and the magnitude of the normal reaction force is $mg\cos\phi$.

### Exercise 3

(i)   What is the total force on the glider in Figure 4?

(ii)  The angle of slope of the table is $\phi = 0.0061$ radians. What is the acceleration vector of the glider?

[*Solution on p. 48*]

The normal reaction is not the only force that a surface can exert on a particle. You know from experience that the effects of *friction* may prevent or inhibit motion of the particle across the surface.

For example, Figure 6 shows a particle resting on a horizontal surface: the forces $\mathbf{F}_g$ and $\mathbf{F}_N$ cancel out. Now suppose that a small horizontal force, $\mathbf{F}_a$, is applied to the particle but that this force is insufficient to get motion started. How can that happen?

The answer can only be that there is another force, $\mathbf{F}_f$, which opposes the motion of the particle. This is called the **force of friction**, or the **frictional reaction**, and its value is such that

$$\mathbf{F}_a + \mathbf{F}_f = \mathbf{0}.$$

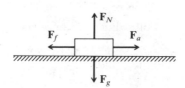

Figure 6

It is clear that the frictional reaction is zero when there is no applied horizontal force. As the applied force increases, the frictional reaction also increases, so that it still cancels the applied force. Eventually, however, the frictional reaction reaches its maximum possible magnitude; after this, the applied force overcomes friction and motion begins. Careful experiments show that friction obeys the following approximate rules:

---

**Properties of friction**

1.  The frictional reaction, $\mathbf{F}_f$, on an object in contact with a solid surface acts parallel to the surface in a direction that opposes motion.

2.  For a static object the magnitude of the frictional reaction is just sufficient to prevent motion; however, it cannot exceed a certain maximum value, which I denote by $|\mathbf{F}_f|^{max}$.

    $|\mathbf{F}_f|^{max}$ is proportional to the magnitude of the normal reaction. We write

    $$|\mathbf{F}_f| \leqslant |\mathbf{F}_f|^{max} = \mu|\mathbf{F}_N|$$

    where the proportionality constant, $\mu$, is known as the **coefficient of static friction**.

3.  For a moving object, the magnitude of the frictional reaction is proportional to the magnitude of the normal reaction. We write

    $$|\mathbf{F}_f| = \mu'|\mathbf{F}_N|$$

    where the proportionality constant, $\mu'$, is known as the **coefficient of kinetic friction**, and is slightly less than $\mu$.

4.  The values of $\mu$ and $\mu'$ depend on the composition and dryness of the surface and of the object resting on it. *They are largely independent of other factors.* (For example, they do not depend on the shape of the object or the area that is in contact with the surface. $\mu$ does not depend on the length of time the object has been resting on the surface and $\mu'$ does not depend on the speed of the object over the surface. Surprisingly, even the roughness of the surface has little influence on $\mu$ and $\mu'$, although *lubrication* can have an important effect.)

---

We can now use the properties of the normal reaction and the frictional reaction to determine whether an object can remain permanently at rest. The following example illustrates how to do this.

### Example 3

A brick rests on a sloping table, as in Figure 7. The coefficient of static friction between the brick and the table is $\mu = 1.0$. What is the maximum value of $\phi$ for which the brick can remain static?

*Solution*

Figure 8 shows my choice of co-ordinate system (the same as in Example 2) and the three forces that act on the brick: gravity, the normal reaction *and* friction.

Following the same analysis as in Example 2, I find that

$$\mathbf{F}_g = mg(\sin\phi\,\mathbf{i} - \cos\phi\,\mathbf{j})$$
$$\mathbf{F}_N = |\mathbf{F}_N|\mathbf{j},$$

and I also find

$$\mathbf{F}_f = |\mathbf{F}_f|(-\mathbf{i}).$$

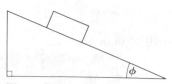

Figure 7

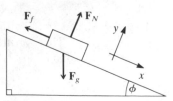

Figure 8

Thus, the total force acting on the brick is

$$\mathbf{F} = \mathbf{F}_g + \mathbf{F}_N + \mathbf{F}_f$$

$$= (mg\sin\phi - |\mathbf{F}_f|)\mathbf{i} + (|\mathbf{F}_N| - mg\cos\phi)\mathbf{j}. \tag{3}$$

In order for the brick to remain static, $\mathbf{F}$ must be equal to the zero vector, so

$$|\mathbf{F}_f| = mg\sin\phi$$

and      $|\mathbf{F}_N| = mg\cos\phi.$

But $|\mathbf{F}_f|$ can only reach a maximum value of $\mu|\mathbf{F}_N|$ so we must have

$$|\mathbf{F}_f| \leqslant \mu|\mathbf{F}_N|;$$

so

$$mg\sin\phi \leqslant \mu mg\cos\phi$$

giving

$$\tan\phi \leqslant \mu = 1.0.$$

Hence $\phi \leqslant \pi/4$ radians and the maximum value of $\phi$ for which the brick can remain static is $\pi/4$ radians.

### Exercise 4

Suppose that the angle $\phi$ in Example 3 is imperceptibly greater than $\pi/4$ radians, so that the brick starts to slide directly down the slope. What is the magnitude of the brick's acceleration if the coefficient of kinetic friction is $\mu' = 0.9$?

[Solution on p. 49]

### Exercise 5

A brick of mass 0.8 kg rests on a table which slopes at $\pi/4$ radians to the horizontal. The coefficient of static friction between the brick and the table is $\mu = 0.5$. An extra force, $\mathbf{F}_1$, is applied to the brick, in a direction perpendicular to the table (see Figure 9). What is the minimum magnitude $|\mathbf{F}_1|$ for which the brick will remain static?

[Solution on p. 49]

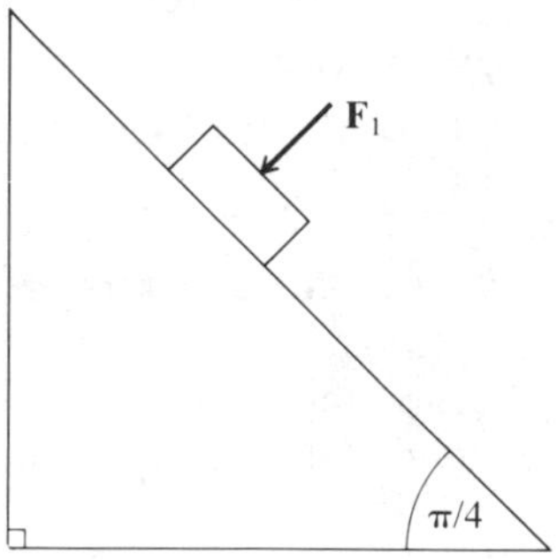

Figure 9

### Summary of Section 3

1.  A **static** particle experiences zero total force.

2.  A particle in contact with a solid surface experiences a **normal reaction** force, $\mathbf{F}_N$, and a **frictional reaction** force, $\mathbf{F}_f$.

3.  $\mathbf{F}_N$ is perpendicular to the surface and $|\mathbf{F}_N|$ is just sufficient to prevent the particle from sinking into the surface.

4.  $\mathbf{F}_f$ is parallel to the surface, in the direction that opposes motion. For a static particle, $|\mathbf{F}_f|$ is just sufficient to prevent sliding over the surface, but the particle can only remain static so long as $|\mathbf{F}_f| \leqslant \mu|\mathbf{F}_N|$, where $\mu$ is the **coefficient of static friction**. For a moving particle, $|\mathbf{F}_f| = \mu'|\mathbf{F}_N|$, where $\mu'$ is the **coefficient of kinetic friction**.

# 4    Uncoupled differential equations and shot-putters (Television Section)

We now turn to problems in which Newton's second law leads to three uncoupled differential equations — one equation for $x$, one for $y$ and one for $z$. This type of problem will be illustrated by the motion of a shot, from the time it leaves the shot-putter's hand to the time it lands on the ground. I shall present this as an example of mathematical modelling, using some of the methods discussed in the *Project Guide*.

## 4.1 The shot-putter's problem

Figure 1 illustrates the motion we shall discuss. It shows the point, $A$, where the shot is released and the point, $B$, where it lands on the ground. In order for the putt to count, the shot-putter must not step beyond the front of the circle, at $O$. Clearly his or her problem is to launch the shot in such a way that the range $OB$ is as long as possible.

Figure 1

Just as in the *Project Guide*, we can construct a feature list of the factors that might influence the path, and range, of the shot. My list is shown below:

1. speed of release
2. angle of release
3. position of release ($A$)
4. mass of shot
5. material of shot
6. diameter of shot
7. spin of shot
8. gravity
9. air resistance
10. weather conditions
11. earth's magnetic field
12. electric and magnetic properties of shot.

Not all these factors are important, and it would be unwise to attempt to take them all into account from the outset. Instead I shall carefully yet drastically prune the list in order to obtain a first, very simple, model. Refinements can be added later, as our understanding grows.

## 4.2 A first model

Pruning the feature list as much as I dare, I am left with five factors to consider:

1. speed of release
2. angle of release
3. position of release
4. mass of shot
5. gravity

I shall assume that all other factors are negligible or irrelevant, so that my model is based on the fiction that the shot moves as a particle and experiences only the downward force of gravity.

In search of simplicity, I shall make two further assumptions about the way the shot is released.

(i)   I shall not worry about the exact position of release of the shot, but will pretend that the shot is released at ground level, from $O$.

(ii)  The shot-putter can launch the shot at any speed up to a maximum value, $v_{max}$, determined by his strength. I shall assume that he can impart the same speed, $v_{max}$, for *any* angle of release.

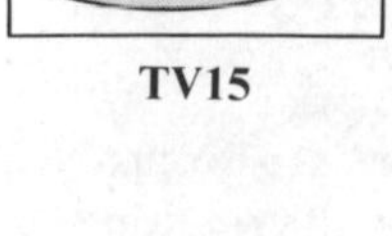

*With these assumptions understood, you are ready to watch the television programme.*

**TV15**

### Summary of programme

The programme used the model I have just described to sketch a solution to the shot-putter's problem. For future reference, the full argument is given below, starting with the diagram in Figure 2.

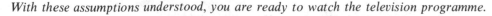

Here, the origin is at the point $O$, at the front of the shot-putter's circle, and the $y$-axis points vertically upwards. At $t = 0$ the shot is launched from $O$. Its initial speed is $v_0$, and its initial direction of motion is in the $x,y$-plane, at $\theta$ radians to the horizontal $x$-axis. The initial conditions are therefore

$$\mathbf{r}(0) = \mathbf{0} \tag{1}$$

and $\quad \dot{\mathbf{r}}(0) = v_0(\cos\theta\,\mathbf{i} + \sin\theta\,\mathbf{j}). \tag{2}$

Newton's second law, applied to Figure 2, gives

$$m\ddot{\mathbf{r}} = \mathbf{F} = \mathbf{F}_g$$

or $\quad m(\ddot{x}\mathbf{i} + \ddot{y}\mathbf{j} + \ddot{z}\mathbf{k}) = mg(-\mathbf{j}) \tag{3}$

which is equivalent to the three *uncoupled* differential equations

$$\ddot{x} = 0, \tag{4a}$$
$$\ddot{y} = -g, \tag{4b}$$
and $\quad \ddot{z} = 0. \tag{4c}$

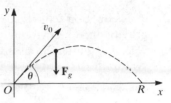

Figure 2

For visual reasons the subscript in $v_0$ was omitted in the TV programme.

These equations can be solved *one at a time* to give

$$x(t) = At + B,$$
$$y(t) = -\tfrac{1}{2}gt^2 + Ct + D,$$
$$z(t) = Gt + H,$$

where $A$, $B$, $C$, $D$, $G$ and $H$ are arbitrary constants. Using the given initial conditions (Equations (1) and (2)) we conclude that $A = v_0\cos\theta$, $B = 0$, $C = v_0\sin\theta$, $D = 0$, $G = 0$ and $H = 0$ so that

$$x(t) = (v_0\cos\theta)t, \tag{5a}$$
$$y(t) = -\tfrac{1}{2}gt^2 + (v_0\sin\theta)t, \tag{5b}$$
$$z(t) = 0. \tag{5c}$$

The shot-putter is really interested in the **path** of the shot — that is, the set of points through which the shot passes. From Equation (5c), the path lies in the $x,y$-plane; its precise shape is found by eliminating $t$ from Equations (5a) and (5b). We find

$$
\begin{aligned}
y &= -\tfrac{1}{2}g\left(\frac{x}{v_0\cos\theta}\right)^2 + v_0\sin\theta\left(\frac{x}{v_0\cos\theta}\right) \\
&= -\left(\frac{g}{2v_0^2\cos^2\theta}\right)x^2 + (\tan\theta)x. \tag{6}
\end{aligned}
$$

Any curve like this, with $y$ a quadratic function of $x$, is known as a **parabola**. Parabolic paths are characteristic of motion under a constant force like gravity.

A shot-putter is especially interested in the point where the shot lands. The $y$-co-ordinate of this point is zero, and its $x$-co-ordinate is denoted by $R$; substituting these values into Equation (6), we obtain

$$0 = -\frac{g}{2v_0^2\cos^2\theta}R^2 + (\tan\theta)R. \tag{7}$$

Thus, either $R = 0$, or

$$
\begin{aligned}
R &= \frac{2v_0^2\cos^2\theta\,\tan\theta}{g} \\
&= \frac{2v_0^2}{g}\cos\theta\sin\theta = \frac{v_0^2}{g}\sin2\theta.
\end{aligned}
$$

The first possibility, $R = 0$, corresponds to the launching of the shot rather than its landing, so we reject this solution and conclude that

$$R = \frac{v_0^2}{g}\sin2\theta. \tag{8}$$

This equation tells us the range of the shot in terms of the acceleration due to gravity and the two quantities, $v_0$ and $\theta$, that describe how the shot is launched. The shot-putter chooses these quantities to obtain the greatest possible value of $R$. Naturally, he uses his full strength, so $v_0 = v_{\text{max}}$ and

$$R = \frac{v_{\text{max}}^2}{g}\sin 2\theta.$$

As shown in Figure 3, $R$ depends on $\theta$, with the maximum range occurring when

$$0 = \frac{dR}{d\theta} = \frac{v_{\text{max}}^2}{g} \times 2\cos 2\theta \qquad \text{i.e. when } 2\theta = \frac{\pi}{2}.$$

So the optimum angle of release is $\theta = \pi/4$ radians (i.e. 45°) and this produces the maximum range

$$R_{\text{max}} = \frac{v_{\text{max}}^2}{g}\sin\frac{\pi}{2}$$

$$= \frac{v_{\text{max}}^2}{g}. \qquad\qquad (9)$$

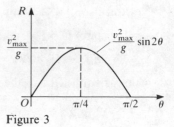

Figure 3

### Diluting gravity

The programme interwove the above analysis with some demonstration experiments. Since the shot is only airborne for about two seconds, we found it more convenient to consider a puck on the surface of a frictionless sloping table. The puck was observed to have a constant acceleration of $0.06\,\text{m}\,\text{s}^{-2}$ down the slope of the table (corresponding to an angle of slope of 0.0061 radians — see Exercise 3 of Section 3). Thus the above analysis should apply to the puck, provided we replace $g$ by 0.06 and choose our $x$- and $y$-axes to lie in the plane of the table, with the $x$-axis horizontal, as in Figure 4. We first checked that the path was parabolic by launching a puck from the origin and finding that the curve $y = -1.0x^2 + 1.6x$ fitted the path quite well.

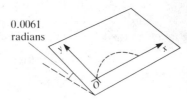

Figure 4

### Exercise 1

Use the data given above to find the initial speed of the puck and the initial angle between its direction of motion and the $x$-axis.

[*Solution on p. 49*]

By looking at the puck's motion from the plane of the table we were able to examine the behaviour of the $x$- and $y$-components separately. This agreed with Equations (4a) and (4b) and so gave direct support for Newton's second law in vector form.

Finally, Equation (8) was checked by placing a target one metre away from the origin along the $x$-axis and hitting it with a puck launched from the origin at $\frac{1}{3}\text{m}\,\text{s}^{-1}$. According to Equation (8) the initial angle of release, $\theta$, satisfies

$$\sin 2\theta = \frac{0.06 \times 1}{(\frac{1}{3})^2} = 0.54,$$

so that $\theta \simeq 16$ degrees (0.29 radians)

or $\qquad \theta \simeq 74$ degrees (1.29 radians).

These angles were confirmed experimentally.

These experiments give us some confidence that our analysis of the first model is correct. The next exercise asks you to use it to answer a question of some interest to shot-putters.

### Exercise 2

In 1981 the world record for the range of a shot-putt was 22.15 metres.

(i)     Estimate the speed of release.

(ii)    Assuming that the floor could take the strain, could expert shot-putters practice inside a gymnasium 30 metres long with a ceiling height of 5 metres?

[*Solution on p. 49*]

## 4.3   A second model: taking height into account

Exercise 2(ii) shows that even world champions only give the shot a maximum height of 5 or 6 metres. This is about three times the shot-putter's height, so Figure 5 is roughly to scale.

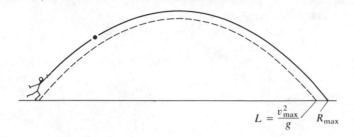

$$L = \frac{v_{max}^2}{g} \quad R_{max}$$

Figure 5

This suggests that the initial height of launch plays a small *but significant* role in determining the maximum range of the shot and the strategy for achieving it: it is no longer clear that the optimum angle is 45°. I shall therefore modify my first model by supposing that the point of release of the shot is

$$\mathbf{r}(0) = h\mathbf{j} \tag{10}$$

where $h \simeq 2$ is the height of the shot-putter in metres. (Equation (10) neglects the possibility of the shot-putter leaning forward beyond the front of the throwing circle, but the distance he can gain in this way is clearly rather small.)

Figure 6 shows my new diagram. Everything is as before, except that the initial value of the $y$-co-ordinate is now $h$ rather than zero. This has the effect of adding $h$ to the right-hand side of Equations (5b), (6) and (7), so the relationship between launch angle and range is now

$$0 = -\frac{g}{2v_0^2\cos^2\theta} R^2 + (\tan\theta)R + h. \tag{11}$$

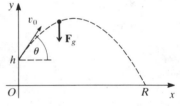

Figure 6

Since I am interested in the *maximum* range, I shall assume that the shot is launched at maximum speed so that $v_0 = v_{max}$. The quantity $\frac{v_{max}^2}{g}$ can be recognized as the maximum range for zero launch height: I shall denote it by the symbol $L$, so Equation (11) becomes

$$0 = -\frac{1}{2L\cos^2\theta} R^2 + (\tan\theta)R + h.$$

The remainder of the calculation is simplified if I express everything in terms of $\tan\theta$. Using

$$\frac{1}{\cos^2\theta} = \sec^2\theta = (1 + \tan^2\theta),$$

I find

$$0 = -\frac{R^2}{2L}(1 + \tan^2\theta) + R\tan\theta + h. \tag{12}$$

The most obvious way of finding the optimum launch angle is to solve Equation (12) for $R$ and then set $\frac{dR}{d\theta}$ equal to zero. However, this leads to tedious algebra which can be avoided by differentiating Equation (12) as it stands. I find

$$0 = -\frac{1}{2L}\left(2R\frac{dR}{d\theta}\right)(1 + \tan^2\theta) - \frac{R^2}{2L}(2\tan\theta\sec^2\theta) + \frac{dR}{d\theta}\tan\theta + R\sec^2\theta.$$

Remember

$$\frac{d}{d\theta}\tan\theta = \sec^2\theta$$

The maximum range, $R_{max}$, is determined by setting $\frac{dR}{d\theta} = 0$, so

$$0 = R_{max}\sec^2\theta\left(-\frac{R_{max}}{L}\tan\theta + 1\right).$$

Rejecting the irrelevant solution, $R_{max} = 0$, and remembering that $\sec^2\theta$ cannot be zero, we conclude that the optimum angle is given by

$$\tan\theta = \frac{L}{R_{max}}. \tag{13}$$

This simple equation reveals an interesting fact. It is clear from Figure 5 that $R_{max} > L$, so it follows that $\tan\theta < 1$ and $\theta < \pi/4$ radians. The exact value is found by combining Equations (12) and (13).

### Exercise 3

(i)    Show that the maximum range is given by

$$R_{max} = L\sqrt{1 + \frac{2h}{L}}. \tag{14}$$

(ii)   Show that the optimum angle, needed to obtain this maximum range, is given by

$$\tan\theta = \frac{1}{\sqrt{1 + \dfrac{2h}{L}}}. \tag{15}$$

[*Solution on p. 49*]

### Exercise 4 (Optional)

Show that your answer to Exercise 3(ii) is equivalent to the statement that the optimum direction of release bisects the angle $CAB$ between the vertical and the straight line from the point of release to the furthest possible point of impact (see Figure 7).

[*Solution on p. 50*]

### Exercise 5

(i)    What is the optimum angle of release for a world champion who launches from a height of 2 metres and achieves a maximum range of 22.15 metres?

(ii)   When I tried to put a shot, I launched it from a height of 1.5 metres, at an angle of 45°, and achieved a range of 5 metres (sic). Estimate my maximum range and the optimum angle for achieving it.

[*Solution on p. 50*]

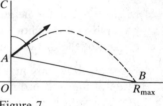

Figure 7

## 4.4   A third model?

It is clear from our initial feature list that our models are not completely realistic. To see whether a third model is needed it is worthwhile making some 'back of the envelope' estimates.

For example, consider the force of air resistance. According to *Unit 4*, this has a magnitude

$$|\mathbf{F}_R| = 0.2\,D^2|\mathbf{v}|^2$$

where $D$ is the diameter of the shot and $|\mathbf{v}|$ is its speed.

Comparing this with the force of gravity on the shot,

$$\frac{|\mathbf{F}_R|}{|\mathbf{F}_g|} = \frac{0.2D^2|\mathbf{v}|^2}{mg}.$$

In SI units a standard shot has $D = 0.15$ and $m = 7.26$. Also, as you saw in Exercise 2(i), $|\mathbf{v}|$ can range up to about 15. We conclude that

$$\frac{|\mathbf{F}_R|}{|\mathbf{F}_g|} \simeq \frac{0.2 \times (0.15)^2 \times (15)^2}{7.26 \times 9.8} \simeq 0.014 \simeq \frac{1}{70}.$$

So the force of air resistance is at least 70 times smaller than the force of gravity, and it should be quite a good approximation to ignore it completely.

Similarly, we could see whether the earth's magnetic field is likely to have an important effect. For example, if the shot carries an electric charge $q$ in a magnetic field $\mathbf{B}$ it will experience a force $\mathbf{F}_B = q(\mathbf{v} \times \mathbf{B})$ (see Example 3 of Section 2). Inserting reasonable values for $q$, $\mathbf{v}$ and $\mathbf{B}$, I conclude that $|\mathbf{F}_B| < 10^{-12}$ newtons, which is certainly negligible!

A final worry concerns our assumption that the shot-putter can impart the same maximum speed to the shot for all angles of release. This is difficult to quantify because it is something that varies from person to person and because serious shot-putters train themselves to impart the highest velocities for angles close to the ones we have calculated. I shall therefore choose this point to draw our modelling of the shot-putter's problem to a close. As with all real problems, this one is open-ended and there are still many questions to ask. For example, what style should be adopted in the initial wind-up before the shot is released? However, this unit is not a manual for shot-putters, so I shall simply leave these questions to one side. You may like to think about them for yourself.

## Summary of Section 4

1.  We modelled the shot as a particle that experiences only the downward force of gravity and assumed that the shot-putter imparts the same maximum speed, $v_{max}$, to the shot for all angles of release.

2.  Newton's second law applied to this problem leads to three uncoupled differential equations which can be solved one by one.

3.  The **path** of the shot is a **parabola**.

4.  If the shot-putter launches from zero height the **maximum range** is

    $$R_{max} = \frac{v_{max}^2}{g} = L$$

    and this is achieved for an angle of release of $\frac{\pi}{4}$ radians.

5.  If the shot-putter launches from a height $h$, the maximum range is

    $$R_{max} = L\sqrt{1 + \frac{2h}{L}}$$

    and this is achieved for an angle of release of

    $$\arctan\left(\frac{1}{\sqrt{1 + \dfrac{2h}{L}}}\right).$$

# 5    Coupled differential equations and pendulums

This section looks at problems in which Newton's second law leads to *coupled* differential equations. Such problems are very common, but I shall restrict myself to one system.

## 5.1    A particle tethered by an inextensible string

Figure 1 shows the system I shall discuss. It consists of a particle, of mass $m$, which is tethered to a fixed point, $O$, by a piece of string. The string is supposed to be **light, inextensible** and **taut**. This means that

(i)    the mass of the string is negligible;

(ii)   the length of the string is a constant, $l$;

(iii)  the string is never slack, so the particle's distance from $O$ is also $l$.

In Figure 2, I have chosen a co-ordinate system whose origin is at the fixed point, $O$, and whose $x$-axis points vertically downwards. This choice leads to a simple description of the forces acting on the particle.

If I neglect air resistance, there are two forces to consider: the force due to gravity, and the force due to the taut string.

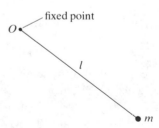

Figure 1

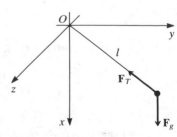

Figure 2

The force due to gravity is denoted by $\mathbf{F}_g$; it acts downwards and has magnitude $mg$ so

$$\mathbf{F}_g = mg\,\mathbf{i}. \tag{1}$$

The force due to the taut string is denoted by $\mathbf{F}_T$; I shall assume that this force acts in the direction shown in Figure 2, from the particle towards the fixed point, $O$. Thus,

$$\hat{\mathbf{F}}_T = -\hat{\mathbf{r}} = -\frac{\mathbf{r}}{|\mathbf{r}|} = -\frac{(x\mathbf{i} + y\mathbf{j} + z\mathbf{k})}{l}$$

where $\mathbf{r}$ is the position vector of the particle and $x$, $y$ and $z$ are its co-ordinates.

The magnitude of $\mathbf{F}_T$ is called the **tension** of the string and is denoted by the symbol $T$, so

Notice that here $T$ does not denote period.

$$|\mathbf{F}_T| = T$$

and $\quad \mathbf{F}_T = -\dfrac{T}{l}(x\mathbf{i} + y\mathbf{j} + z\mathbf{k}). \tag{2}$

Combining Equations (1) and (2), the total force acting on the particle is

$$\mathbf{F} = \mathbf{F}_g + \mathbf{F}_T$$

$$= \left(mg - \frac{T}{l}x\right)\mathbf{i} - \frac{T}{l}y\mathbf{j} - \frac{T}{l}z\mathbf{k}$$

and Newton's second law gives

$$m\ddot{x} = mg - \frac{T}{l}x; \tag{3a}$$

$$m\ddot{y} = -\frac{T}{l}y; \tag{3b}$$

$$m\ddot{z} = -\frac{T}{l}z. \tag{3c}$$

**Exercise 1**

Suppose that the particle moves in a horizontal circle, so that $x = x_0$, a constant.

(i)     Find the tension in the string.

(ii)    Solve Equations (3b) and (3c) and hence show that the time taken to complete one revolution is $2\pi\sqrt{\dfrac{x_0}{g}}$.

[Solution on p. 50]

At first sight, Equations (3a)–(3c) look very similar to Equations (6a)–(6c) of Section 2, with $\dfrac{T}{l}$ replacing the constant $c$. In the case studied in Exercise 1, $\dfrac{T}{l}$ is a constant and the analogy is complete. However, this is quite exceptional. In general, the tension will *not* be constant, but will depend on some combination of $x$, $y$, $z$, $\dot{x}$, $\dot{y}$ and $\dot{z}$. When this happens, Equations (3a)–(3c) are *coupled together* and we must find a way of unravelling them. The next subsection will show how this is done for motion in the vertical $x,y$-plane.

## 5.2   The motion of a pendulum bob

It is clear that Equation (3c) has a particular solution of the form

$$z(t) = 0. \tag{4}$$

We can therefore consider a class of motions in which the particle never strays from the $x,y$-plane. Since the string is assumed to be taut the particle will move, like a pendulum bob, on a vertical circle of radius $l$. This is illustrated in Figure 3.

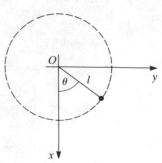

Figure 3

The $x$- and $y$-components of the pendulum bob satisfy Equations (3a) and (3b), but in order to solve these equations we must overcome two difficulties:

(i)    the tension $T$ depends on $x$, $y$, $\dot{x}$ and $\dot{y}$ in some unknown way;

(ii)   the equations are coupled together.

The first difficulty is best avoided by eliminating $T$ from our equations. This can be done by multiplying Equation (3a) by $y$, and multiplying Equation (3b) by $x$, to obtain

$$my\ddot{x} = mgy - \frac{T}{l}yx;$$

$$mx\ddot{y} = \qquad -\frac{T}{l}xy.$$

Then, subtracting the first of these equations from the second, I obtain

$$mx\ddot{y} - my\ddot{x} = -mgy. \tag{5}$$

Equation (5) has the merit of describing the motion of the pendulum bob without referring to the unknown tension. The only snag is that it involves both $x$ and $y$ and so cannot be solved by the techniques of *Unit 6*. Fortunately, we have the additional information that the string has constant length. This suggests that we should express the Cartesian co-ordinates ($x$ and $y$) in terms of polar co-ordinates ($l$ and $\theta$). Then, since $l$ remains constant, and only $\theta$ varies as the bob moves round the circle, we should be able to rewrite Equation (5) as a differential equation for $\theta$.

Figure 4 shows the relationship between Cartesian co-ordinates and polar co-ordinates. We have

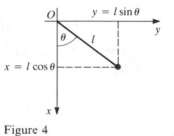

Figure 4

$$x = l\cos\theta; \tag{6a}$$
$$y = l\sin\theta. \tag{6b}$$

Since $l$ is a constant, these equations give

$$\dot{x} = -l\sin\theta\,\dot{\theta}; \tag{7a}$$
$$\dot{y} = l\cos\theta\,\dot{\theta}. \tag{7b}$$

Differentiating again,

$$\ddot{x} = -l\cos\theta\,\dot{\theta}^2 - l\sin\theta\,\ddot{\theta}; \tag{8a}$$
$$\ddot{y} = -l\sin\theta\,\dot{\theta}^2 + l\cos\theta\,\ddot{\theta}. \tag{8b}$$

Finally, substituting for $x$, $y$, $\ddot{x}$ and $\ddot{y}$ in Equation (5), I obtain

$$ml\cos\theta(-l\sin\theta\,\dot{\theta}^2 + l\cos\theta\,\ddot{\theta}) - ml\sin\theta(-l\cos\theta\,\dot{\theta}^2 - l\sin\theta\,\ddot{\theta})$$
$$= -mgl\sin\theta.$$

Collecting terms in $\dot{\theta}^2$ and $\ddot{\theta}$, this becomes

$$ml^2\dot{\theta}^2(-\cos\theta\sin\theta + \sin\theta\cos\theta) + ml^2\ddot{\theta}(\cos^2\theta + \sin^2\theta)$$
$$= -mgl\sin\theta.$$

So, using $\cos^2\theta + \sin^2\theta = 1$, and dividing by $ml^2$,

$$\ddot{\theta} = -\frac{g}{l}\sin\theta. \tag{9}$$

This is the equation we have been looking for — it involves just *one* variable, $\theta$, and its second derivative, $\ddot{\theta}$, so we have succeeded in uncoupling Equations (3a) and (3b). If we could find the function $\theta(t)$ that satisfies Equation (9), subject to appropriate initial conditions, we could use Equations (4), (6a) and (6b) to find the position vector at any later time, $t$:

$$\mathbf{r}(t) = l\cos\theta(t)\mathbf{i} + l\sin\theta(t)\mathbf{j}.$$

I shall not solve Equation (9) here — that would be too difficult a task; instead I shall make a connection between our present discussion of a pendulum and that given in *Unit 7*.

In *Unit 7* we claimed that the total mechanical energy of a circular pendulum is

$$E = \tfrac{1}{2}m(l\dot\theta)^2 + mgl(1 - \cos\theta). \tag{10}$$

The first term is the kinetic energy of the bob, $\tfrac{1}{2}$ mass × (speed)$^2$. You can see that this agrees with Equations (4), (7a) and (7b) because

$$\begin{aligned}
\tfrac{1}{2}m|\dot{\mathbf{r}}|^2 &= \tfrac{1}{2}m(\dot{x}^2 + \dot{y}^2 + \dot{z}^2) \\
&= \tfrac{1}{2}m(l^2\sin^2\theta\,\dot\theta^2 + l^2\cos^2\theta\,\dot\theta^2) \\
&= \tfrac{1}{2}ml^2\dot\theta^2.
\end{aligned}$$

The second term is the gravitational potential energy of the bob, mass × g × height. This agrees with Equation (6a) because

$$\begin{aligned}
mg \times (\text{height above } x = l) &= mg(l - x) \\
&= mg(l - l\cos\theta).
\end{aligned}$$

Now, differentiating Equation (10), I find that

$$\frac{dE}{dt} = ml^2\dot\theta\,\ddot\theta + mgl\sin\theta\,\dot\theta$$

$$= ml^2\dot\theta\left(\ddot\theta + \frac{g}{l}\sin\theta\right)$$

so, using Equation (9),

$$\frac{dE}{dt} = 0.$$

Thus, $E$ remains constant in time, and I conclude that the total energy, defined by Equation (10), is conserved — just as we assumed in *Unit 7*!

In view of the close connection between Equations (9) and (10), it is natural to ask what progress we have made since *Unit 7*. I think there are three points to consider.

Firstly, we have gained rigour. In *Unit 7* we merely *assumed* that the law of conservation of mechanical energy applies to a pendulum. Now we have *proved* this fact, directly from Newton's second law.

Secondly, we have set up a formalism that is more general than the law of conservation of mechanical energy. Newton's second law, in vector form, can be used even when friction and air resistance prevent us from using energy conservation.

Thirdly, Equations (3a) and (3b) contain more information than is provided by Equation (10), because they tell us about the tension in the string. This information was lost when I eliminated $T$ to obtain Equation (5); it will be recovered in the next subsection.

## 5.3   The tension of a pendulum string

In order to find the tension in the string I must go back to Equations (3a) and (3b):

$$m\ddot{x} = mg - \frac{T}{l}x;$$

$$m\ddot{y} = \quad\ - \frac{T}{l}y.$$

Using Equations (3a) and (3b), (8a) and (8b) to write these equations in terms of polar co-ordinates, I have

$$m(-l\cos\theta\,\dot\theta^2 - l\sin\theta\,\ddot\theta) = mg - T\cos\theta; \tag{11a}$$

$$m(-l\sin\theta\,\dot\theta^2 + l\cos\theta\,\ddot\theta) = \quad\ -T\sin\theta \tag{11b}$$

The tension can now be found by multiplying Equation (11a) by $\cos\theta$, multiplying Equation (11b) by $\sin\theta$, and adding the two resulting equations. This gives

$$ml(-\cos^2\theta\,\dot\theta^2 - \cos\theta\sin\theta\,\ddot\theta - \sin^2\theta\,\dot\theta^2 + \sin\theta\cos\theta\,\ddot\theta)$$
$$= mg\cos\theta - T(\cos^2\theta + \sin^2\theta);$$

using $\cos^2\theta + \sin^2\theta = 1$, this becomes

$$-ml\dot\theta^2 = mg\cos\theta - T$$

so that

$$T = mg\cos\theta + ml\dot\theta^2 \tag{12}$$

or, in terms of Cartesian co-ordinates,

$$\cos\theta = \frac{x}{l} \quad\text{and}\quad \dot\theta^2 = \frac{\dot x^2 + \dot y^2}{l^2},$$

$$T = \frac{m}{l}(gx + \dot x^2 + \dot y^2). \tag{13}$$

Notice that the tension is *not* constant, but depends on the height and speed of the bob. This should not surprise you. If the tension *were* constant, Equations (3a)–(3c) would give rise to simple harmonic motion with a period that was *independent of amplitude*. But you already know that a pendulum does not oscillate simple harmonically — its period *increases* with increasing amplitude. This is a consequence of the changing tension in the string.

*See Section 3 of Unit 7.*

### Exercise 2: Why strong thread is needed for conkers

(i)   Combine Equations (10) and (12) to show that the tension in the string of a pendulum varies as

$$T = mg(3\cos\theta - 2) + \frac{2E}{l} \tag{14}$$

where $m$ is the mass of the bob

$l$ is the length of the string

and $E$ is the total energy, with the zero of potential energy at the lowest point of the bob's path.

(ii)   Suppose that the pendulum bob is released from rest when $\theta = \pi/2$. Show that, if the string is not to break during the subsequent oscillations, it must be able to sustain a tension of at least $3mg$.

(iii)   The bob is released from rest when $\theta = \pi/2$, but the string can only sustain a tension of $1.5mg$. At what angle $\theta$ does the string break?

[*Solution on p. 51*]

So far, we have assumed that the string of the pendulum is always taut so that the bob remains a fixed distance $l$ from the origin, and is held on its circular path by the tension in the string. However, Equation (14) shows that $T$ decreases as the angle $\theta$ increases, so there may come a point at which the tension in the string is zero. When this happens, the string exerts no forces on the bob, which moves under gravity alone. The bob then falls closer to the origin and the string goes slack. So long as the string is slack its tension remains equal to zero — remember that strings can pull but cannot push!

### Exercise 3: Looping the loop

A pendulum bob of mass $m$ is supported by a string of length $l$. The bob is propelled horizontally from the bottom of its arc with a speed $v_0$.

(i)   Show that, in order for the bob to describe a complete circle, with the string never going slack, $v_0$ must be greater than $\sqrt{5gl}$.

(ii)   At what angle does the string go slack if $v_0 = 2\sqrt{gl}$?

[*Solution on p. 51*]

*Problem 4 in Section 6 will examine the motion of the bob after the string has gone slack.*

## 5.4 Using energy conservation

The fact that the tension of the string given in Equation (13) depends on velocity raises an interesting question. In one-dimensional mechanics you saw that velocity-dependent forces, like air resistance or viscous damping, prevent us from using the conservation of energy. Now we have shown that a pendulum bob experiences a velocity-dependent force, $\mathbf{F}_T = T(-\hat{\mathbf{r}})$, yet it satisfies the law of conservation of mechanical energy with the sum of its kinetic and potential energies remaining constant. Why does this happen?

The reason is that $\mathbf{F}_T$ acts *radially*, along the line of the string, while the bob moves *tangentially*, round the circle. So $\mathbf{F}_T$ is always perpendicular to $\dot{\mathbf{r}}$, as shown in Figure 5.

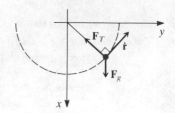

Figure 5

The significance of this fact is revealed by calculating the rate of change of kinetic energy of the bob:

$$\frac{d}{dt}(\tfrac{1}{2}m|\dot{\mathbf{r}}|^2) = \tfrac{1}{2}m\frac{d}{dt}(\dot{x}^2 + \dot{y}^2 + \dot{z}^2)$$
$$= \tfrac{1}{2}m(2\dot{x}\ddot{x} + 2\dot{y}\ddot{y} + 2\dot{z}\ddot{z})$$
$$= m\ddot{\mathbf{r}} \cdot \dot{\mathbf{r}}$$
$$= (\mathbf{F}_T + \mathbf{F}_g) \cdot \dot{\mathbf{r}}, \quad \text{by Newton's second law.}$$

It is here that the relative directions of $\mathbf{F}_T$ and $\dot{\mathbf{r}}$ play their crucial role. Since $\mathbf{F}_T$ is perpendicular to $\dot{\mathbf{r}}$, we have $\mathbf{F}_T \cdot \dot{\mathbf{r}} = 0$ and

$$\frac{d}{dt}(\tfrac{1}{2}m|\dot{\mathbf{r}}|^2) = \mathbf{F}_g \cdot \dot{\mathbf{r}}. \tag{15}$$

So the tension in the string has no influence on the kinetic energy: it affects the direction of motion of the bob, but it does not change the speed. We are left with Equation (15), which involves only gravity, and this is easily rearranged to give

$$\frac{d}{dt}(\tfrac{1}{2}m|\dot{\mathbf{r}}|^2) = mg\mathbf{i} \cdot \dot{\mathbf{r}} = mg\dot{x} = \frac{d}{dt}(mgx)$$

or

$$\frac{d}{dt}(\tfrac{1}{2}m|\dot{\mathbf{r}}|^2 + mg(-x)) = 0 \tag{16}$$

$mg(-x)$ is the gravitational potential energy because $x$ is measured vertically *downwards*.

which tells us that the sum of the kinetic and gravitational potential energies of the bob is conserved, in spite of the velocity-dependent force, $\mathbf{F}_T$, provided by the string.

### Exercise 4

Use the conservation of energy to show that the particle described in Exercise 1 performs *uniform* circular motion.

[*Solution on p. 51*]

Of course, the above argument will work for *any* force that is perpendicular to the particle's velocity so you can now justify the use of Equation (16) for many different systems.

### Exercise 5

A co-ordinate system is chosen with an $x$-axis that points vertically downwards. In which of the following cases would Equation (16) apply? Explain your answers.

(i)   A bead sliding on a curved wire without friction or air resistance.

(ii)  A bead sliding on a curved wire with friction and/or air resistance.

(iii) A charged particle falling under gravity without air resistance but with the additional magnetic force $\mathbf{F}_B = q(\mathbf{v} \times \mathbf{B})$ described in Example 3 of Section 2.

[*Solution on p. 51*]

## Summary of Section 5

1. This section discussed a particle of mass $m$ suspended from a fixed point by a **light taut inextensible** string of length $l$. The angle between the string and the downward vertical is $\theta$.

2. The particle can move in a horizontal circle. In this case its speed is constant, the tension in the string is constant, and the equations of motion for $x$, $y$, and $z$ are uncoupled.

3. The particle can move in a vertical circle. In this case its speed is *not* constant, the tension in the string is *not* constant and the equations of motion are coupled.

   The particle moves in such a way that

   $$\ddot{\theta} = -\frac{g}{l}\sin\theta,$$

   $$\tfrac{1}{2}m(l\dot{\theta})^2 + mgl(1 - \cos\theta) = E = \text{constant}$$

   and the tension in the string is

   $$T = ml\dot{\theta}^2 + mg\cos\theta$$
   $$= mg(3\cos\theta - 2) + \frac{2E}{l},$$

   so long as the right-hand side of this equation is positive. The string goes slack as soon as $T = 0$, and from then onwards a different description of the motion must be used (see Problem 4 in Section 6).

4. The sum of the kinetic and gravitational potential energies of a particle is conserved provided it experiences only the force of gravity and other forces (like the tension in a string, the normal reaction force or the magnetic force on a charged particle) that act at right-angles to the particle's velocity.

# 6   End of unit problems

### Problem 1: An immovable object

A brick of mass $m$ is initially at rest on a horizontal surface. Its upper face is pushed by a force, $\mathbf{F}_1$, which points at an angle $\theta$ radians below the horizontal (see Figure 1). If the coefficient of static friction between the brick and the surface is $\mu$, show that no matter how large $|\mathbf{F}_1|$, the brick will not budge if $\mu > \cot\theta$.

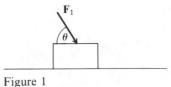

Figure 1

[*Solution on p. 51*]

### Problem 2: Longfellow's tall story

> Swift of foot was Hiawatha;
> He could shoot an arrow from him,
> And run forward with such fleetness,
> That the arrow fell behind him!

> Strong of arm was Hiawatha;
> He could shoot ten arrows upward,
> Shoot them with such strength and swiftness,
> That the tenth had left the bow-string
> Ere the first to earth had fallen!

Assuming that Hiawatha could shoot arrows at the rate of one every second, and that he tried for the maximum range when he 'shot an arrow from him', how swift of foot was he? Treat the arrows as particles, launched from ground level and neglect the force of air resistance.

[*Solution on p. 52*]

**Problem 3: A vital decision**

A motorist finds himself travelling rapidly towards a brick wall on the far side of a T-junction (see Figure 2). Two possible options occur to him:

(a)  to maintain a constant speed and steer the car in the arc of a circle, just avoiding the wall.

(b)  to steer straight at the wall, but to brake as hard as possible.

If the magnitude of the total force on the car is the same in either case, which of the two options would be most likely to avert disaster?

[*Solution on p. 52*]

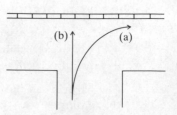

Figure 2

**Problem 4: Almost looping the loop**

A pendulum has a string of length $l$ and a bob of mass $m$. The bob is initially at rest at its static position.

(i)   Suppose the bob is propelled vertically upwards with a speed $2\sqrt{gl}$. Show that it *just* reaches the inverted position of Figure 3, with the string taut and pointing vertically upwards from the pivot.

(ii)  Suppose the bob is propelled sideways from its static position with the speed $2\sqrt{gl}$. It follows a path like that drawn in Figure 4: to begin with the path is a circle, but eventually the string goes slack and the path becomes a parabola as the bob feels only the force of gravity. What is the highest point reached by the bob in this case?

(*Hint:* use the answer to Exercise 3(ii) in Section 5)

[*Solution on p. 53*]

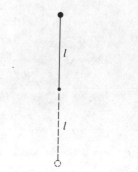

Figure 3

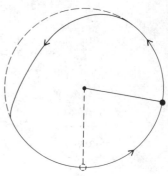

Figure 4

# Appendix 1:  Solutions to the exercises

## Solutions to the exercises in Section 1

**1.(i)**  Using $\mathbf{r}(t) = \pi^2 t^2 \mathbf{i}$, I obtain the following table and diagram:

| $t$ | $\mathbf{r}(t)$ |
|-----|-----------------|
| 0    | **0**        |
| 0.25 | $0.62\,\mathbf{i}$ |
| 0.5  | $2.47\,\mathbf{i}$ |
| 0.75 | $5.55\,\mathbf{i}$ |
| 1    | $9.87\,\mathbf{i}$ |

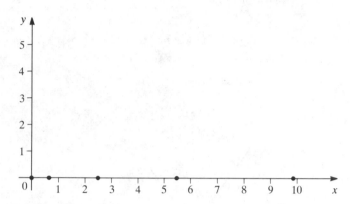

**(ii)** From the diagram, the particle appears to move in a straight line with increasing speed.

**2.(i)**  Using $\mathbf{r}(t) = 2\cos\pi t\,\mathbf{i} + 2\sin\pi t\,\mathbf{j}$, I obtain the following table and diagram:

| $t$ | $\mathbf{r}(t)$ |
|-----|-----------------|
| 0    | $2\,\mathbf{i}$ |
| 0.25 | $1.41\,\mathbf{i} + 1.41\,\mathbf{j}$ |
| 0.5  | $2\,\mathbf{j}$ |
| 0.75 | $-1.41\,\mathbf{i} + 1.41\,\mathbf{j}$ |
| 1    | $-2\,\mathbf{i}$ |
| 1.25 | $-1.41\,\mathbf{i} - 1.41\,\mathbf{j}$ |
| 1.5  | $-2\,\mathbf{j}$ |
| 1.75 | $1.41\,\mathbf{i} - 1.41\,\mathbf{j}$ |

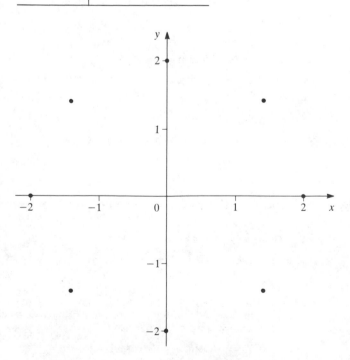

**(ii)** From the diagram, the particle appears to move in a circle with constant speed.

**3.** The second example discussed on the tape had

$$\mathbf{r} = 2\cos\pi t\,\mathbf{i} + 2\sin\pi t\,\mathbf{j};$$

$$|\dot{\mathbf{r}}| = 2\pi; \qquad \frac{d}{dt}|\mathbf{r}| = 0;$$

$$|\ddot{\mathbf{r}}| = 2\pi^2; \qquad \frac{d}{dt}|\dot{\mathbf{r}}| = 0.$$

It therefore illustrates the fact that, in general,

$$|\dot{\mathbf{r}}| \neq \frac{d}{dt}|\mathbf{r}| \qquad \text{and} \qquad |\ddot{\mathbf{r}}| \neq \frac{d}{dt}|\dot{\mathbf{r}}|.$$

**4.**  Yes, at $t = 1$.

According to the tape, the two acceleration vectors are equal when

$$2\pi^2\,\mathbf{i} = -2\pi^2\cos\pi t\,\mathbf{i} - 2\pi^2\sin\pi t\,\mathbf{j}.$$

That is, when $\sin\pi t = 0$ and $\cos\pi t = -1$. Between $t = 0$ and $t = 2$, the only time this happens is at $t = 1$; at this instant the two accelerations are the same in magnitude *and* direction.

**5.**  $\mathbf{r}(t) = 10t\,\mathbf{i} + (2 + 10t - 4.9t^2)\mathbf{j}$

**(i)**  The particle's velocity vector is

$$\mathbf{v}(t) = \frac{d}{dt}(10t\,\mathbf{i} + (2 + 10t - 4.9t^2)\mathbf{j})$$

$$= 10\,\mathbf{i} + (10 - 9.8t)\mathbf{j}.$$

Taking the magnitude of the velocity vector, I obtain the particle's speed:

$$|\mathbf{v}(t)| = \sqrt{10^2 + (10 - 9.8t)^2}.$$

At $t = 0$,

$$|\mathbf{v}(0)| = \sqrt{10^2 + 10^2} \simeq 14.14,$$

so the speed of the particle at $t = 0$ is $14.14\,\text{ms}^{-1}$.

**(ii)**  The particle's direction of motion at $t = 0$ is that of the unit vector

$$\hat{\mathbf{v}}(0) = \frac{\mathbf{v}(0)}{|\mathbf{v}(0)|} \simeq \frac{10\mathbf{i} + 10\mathbf{j}}{14.14} \simeq 0.707\mathbf{i} + 0.707\mathbf{j}.$$

**(iii)**  The particle's acceleration vector at time $t$ is

$$\mathbf{a}(t) = \frac{d}{dt}(10\mathbf{i} + (10 - 9.8t)\mathbf{j})$$

$$= -9.8\mathbf{j}.$$

**6.**  To see whether the particle comes instantaneously to rest I must find out whether $\mathbf{v}$ is ever equal to the zero vector. The question gives the acceleration vector as a function of time:

$$\mathbf{a}(t) = (2t - 1)\mathbf{i} + \mathbf{j} + 2t\mathbf{k}.$$

Integrating each component in turn, and remembering to include arbitrary constants $A$, $B$ and $C$, I find

$$\mathbf{v}(t) = (t^2 - t + A)\mathbf{i} + (t + B)\mathbf{j} + (t^2 + C)\mathbf{k}.$$

The initial condition $\mathbf{v}(0) = -2\mathbf{i} - 2\mathbf{j} - 4\mathbf{k}$ then gives

$$-2\mathbf{i} - 2\mathbf{j} - 4\mathbf{k} = A\mathbf{i} + B\mathbf{j} + C\mathbf{k}$$

so that

$$A = -2, \qquad B = -2, \qquad C = -4$$

and

$$\mathbf{v}(t) = (t^2 - t - 2)\mathbf{i} + (t - 2)\mathbf{j} + (t^2 - 4)\mathbf{k}$$

$$= (t - 2)((t + 1)\mathbf{i} + \mathbf{j} + (t + 2)\mathbf{k}).$$

From this expression, it is clear that $\mathbf{v}(2) = \mathbf{0}$, so that the particle comes instantaneously to rest at $t = 2$.

To find the position of the particle at this instant, I integrate each component of the velocity vector to obtain

$$\mathbf{r}(t) = \left(\frac{t^3}{3} - \frac{t^2}{2} - 2t + D\right)\mathbf{i} + \left(\frac{t^2}{2} - 2t + G\right)\mathbf{j}$$
$$+ \left(\frac{t^3}{3} - 4t + H\right)\mathbf{k}$$

where $D$, $G$ and $H$ are arbitrary constants that are chosen to agree with the given initial condition $\mathbf{r}(0) = \frac{1}{3}\mathbf{i} - \frac{2}{3}\mathbf{k}$. This gives:

$$\tfrac{1}{3}\mathbf{i} - \tfrac{2}{3}\mathbf{k} = D\mathbf{i} + G\mathbf{j} + H\mathbf{k}$$

so that

$$D = \tfrac{1}{3}, \qquad G = 0, \qquad H = -\tfrac{2}{3} \qquad \text{and}$$
$$\mathbf{r}(t) = \left(\frac{t^3}{3} - \frac{t^2}{2} - 2t + \frac{1}{3}\right)\mathbf{i} + \left(\frac{t^2}{2} - 2t\right)\mathbf{j}$$
$$+ \left(\frac{t^3}{3} - 4t - \frac{2}{3}\right)\mathbf{k}.$$

Hence, at $t = 2$,

$$\mathbf{r}(2) = \left(\frac{8}{3} - 2 - 4 + \frac{1}{3}\right)\mathbf{i} + (2 - 4)\mathbf{j} + \left(\frac{8}{3} - 8 - \frac{2}{3}\right)\mathbf{k}$$
$$= -3\mathbf{i} - 2\mathbf{j} - 6\mathbf{k}.$$

The distance of the particle from the origin is then

$$|\mathbf{r}(2)| = \sqrt{(-3)^2 + (-2)^2 + (-6)^2} = 7.$$

I conclude that the particle *does* come instantaneously to rest, 2 seconds after $t = 0$, at a point 7 metres from the origin.

7.  In the diagram below (where the $z$-axis is perpendicular to the page) the angles $\mathbf{F}_g$ makes with the $x$-, $y$- and $z$-axes respectively are:

$$\alpha = \frac{\pi}{4} + \frac{\pi}{2} = \frac{3\pi}{4}, \qquad \beta = \frac{\pi}{4} + \frac{\pi}{2} = \frac{3\pi}{4},$$
$$\gamma = \frac{\pi}{2}.$$

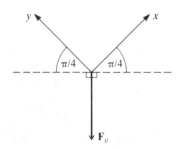

Hence,

$$\cos\alpha \simeq -0.707, \qquad \cos\beta \simeq -0.707, \qquad \cos\gamma = 0,$$

and      $\hat{\mathbf{F}}_g = -0.707\mathbf{i} - 0.707\mathbf{j}.$

The magnitude of the gravitational force on the 3 kg particle is

$$|\mathbf{F}_g| = mg \simeq 3 \times 9.8 = 29.4,$$

so      $\mathbf{F}_g \simeq 29.4(-0.707\mathbf{i} - 0.707\mathbf{j}) \simeq -20.8\mathbf{i} - 20.8\mathbf{j}.$

8.  From Figure 11 in the text,

$$\hat{\mathbf{F}}_3 = \cos(\pi - \phi)\mathbf{i} + \cos\left(\frac{\pi}{2} - \phi\right)\mathbf{j}$$

and      $\hat{\mathbf{F}}_4 = \cos\phi\,\mathbf{i} + \cos\left(\phi + \frac{\pi}{2}\right)\mathbf{j}.$

Using Equations (17), (18) and (19) these expressions simplify to

$$\hat{\mathbf{F}}_3 = -\cos\phi\,\mathbf{i} + \sin\phi\,\mathbf{j};$$
$$\hat{\mathbf{F}}_4 = \cos\phi\,\mathbf{i} - \sin\phi\,\mathbf{j}.$$

(The same result could also be obtained by using the trigonometry shown in the diagrams below.)

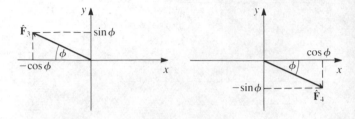

Finally, scaling by the force magnitudes,

$$\mathbf{F}_3 = |\mathbf{F}_3|(-\cos\phi\,\mathbf{i} + \sin\phi\,\mathbf{j});$$
$$\mathbf{F}_4 = |\mathbf{F}_4|(\cos\phi\,\mathbf{i} - \sin\phi\,\mathbf{j}).$$

9.

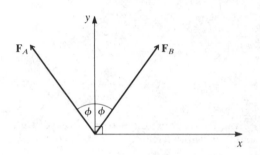

From the diagram above, and using Equation (14), we find

$$\hat{\mathbf{F}}_A = \cos\left(\phi + \frac{\pi}{2}\right)\mathbf{i} + \cos\phi\,\mathbf{j}$$
$$= -\sin\phi\,\mathbf{i} + \cos\phi\,\mathbf{j},$$

and      $\hat{\mathbf{F}}_B = \cos\left(\frac{\pi}{2} - \phi\right)\mathbf{i} + \cos\phi\,\mathbf{j}$
$$= \sin\phi\,\mathbf{i} + \cos\phi\,\mathbf{j}.$$

These results could also be obtained using the trigonometric method of Figure 10.

By Hooke's Law, the force magnitudes are:

$$|\mathbf{F}_A| = |\mathbf{F}_B| = k(l - l_0)$$

so      $\mathbf{F}_A = k(l - l_0)(-\sin\phi\,\mathbf{i} + \cos\phi\,\mathbf{j})$

and      $\mathbf{F}_B = k(l - l_0)(\sin\phi\,\mathbf{i} + \cos\phi\,\mathbf{j}).$

10.(i)   $B, C$;      (ii)   $A, E$;      (iii)   $D, F$.

11.  $\mathbf{F} = \mathbf{F}_A + \mathbf{F}_B + \mathbf{F}_g$
$$= k(l - l_0)(-\sin\phi\,\mathbf{i} + \cos\phi\,\mathbf{j})$$
$$+ k(l - l_0)(\sin\phi\,\mathbf{i} + \cos\phi\,\mathbf{j}) + mg(-\mathbf{j})$$
$$= (2k(l - l_0)\cos\phi - mg)\mathbf{j}.$$

## Solutions to the exercises in Section 2

1.(i)   $\mathbf{r}(t) = l\cos\omega t\,\mathbf{i} + l\sin\omega t\,\mathbf{j}$

and   $|\mathbf{r}(t)| = \sqrt{(l\cos\omega t)^2 + (l\sin\omega t)^2}$
$$= l\sqrt{\cos^2\omega t + \sin^2\omega t} = l.$$

So the particle moves in a plane (the $x,y$-plane) and remains a constant distance, $l$, from the origin; it must therefore move on a circle. The velocity of the particle is

$$\dot{\mathbf{r}}(t) = -l\omega\,\sin\omega t\,\mathbf{i} + l\omega\,\cos\omega t\,\mathbf{j}$$

and
$$|\dot{\mathbf{r}}(t)| = \sqrt{(-l\omega\sin\omega t)^2 + (l\omega\cos\omega t)^2}$$
$$= l\omega\sqrt{\sin^2\omega t + \cos^2\omega t} = l\omega.$$

So the particle has a constant speed, $l\omega$. The time, $T$, taken to complete one orbit is the length of the circumference of the circle, divided by the speed of the particle, so

$$T = \frac{2\pi l}{l\omega} = \frac{2\pi}{\omega}.$$

$\left(\text{This agrees with the observation that } \mathbf{r}\left(t + \frac{2\pi}{\omega}\right) = \mathbf{r}(t).\right)$

**(ii)** According to Example 1, the force acting on the particle is

$$\mathbf{F} = ml\omega^2(-\hat{\mathbf{r}}).$$

This points from the particle towards the centre of the circle, so the spring must be stretched rather than compressed. According to Hooke's Law, the force provided by a stretched perfect spring of length $l$, natural length $l_0$ and stiffness $k$, has magnitude

$$|\mathbf{F}_s| = k(l - l_0);$$

comparing this with the magnitude of $\mathbf{F}$, we have
$$ml\omega^2 = k(l - l_0)$$
so
$$\omega = \sqrt{\frac{k}{m}\left(\frac{l - l_0}{l}\right)}$$

and $T = \dfrac{2\pi}{\omega} = 2\pi\sqrt{\dfrac{m}{k}\left(\dfrac{l}{l - l_0}\right)}.$

However, $l \geqslant l - l_0$

so $\dfrac{l}{l - l_0} \geqslant 1$ for all values of $l$ and so

$$T \geqslant 2\pi\sqrt{\frac{m}{k}}.$$

Hence, the time, $T$, taken to complete a circular orbit is always greater than $2\pi\sqrt{\dfrac{m}{k}}$ (which is the period of straight-line oscillations).

**2.** $\mathbf{r}(t) = \cos\omega t\,\mathbf{r}(0) + \dfrac{1}{\omega}\sin\omega t\,\dot{\mathbf{r}}(0) + \dfrac{g}{\omega^2}(1 - \cos\omega t)\mathbf{i}.$

**(i)** If $\mathbf{r}(0) = \dfrac{g}{\omega^2}\mathbf{i}$ and $\dot{\mathbf{r}}(0) = \mathbf{0}$,

$$\mathbf{r}(t) = \cos\omega t\left(\frac{g}{\omega^2}\mathbf{i}\right) + \frac{g}{\omega^2}(1 - \cos\omega t)\mathbf{i}$$
$$= \frac{g}{\omega^2}\mathbf{i},$$

so the particle remains permanently at rest, at the point $\dfrac{g}{\omega^2}\mathbf{i}$.

If $\mathbf{r}(0) = \left(\dfrac{g}{\omega^2} + A\right)\mathbf{i}$ and $\dot{\mathbf{r}}(0) = \omega B\mathbf{i}$,

$$\mathbf{r}(t) = \cos\omega t\left(\frac{g}{\omega^2} + A\right)\mathbf{i} + \frac{1}{\omega}\sin\omega t(\omega B\mathbf{i})$$

$$+ \frac{g}{\omega^2}(1 - \cos\omega t)\mathbf{i}$$

$$= \left(A\cos\omega t + B\sin\omega t + \frac{g}{\omega^2}\right)\mathbf{i}$$

so the particle oscillates simple-harmonically along the vertical $x$-axis.

**(ii) (a)** If $\mathbf{r}(0) = \dfrac{g}{\omega^2}\mathbf{i} + A\mathbf{j}$ and $\dot{\mathbf{r}}(0) = \omega B\mathbf{j}$,

$$\mathbf{r}(t) = \cos\omega t\left(\frac{g}{\omega^2}\mathbf{i} + A\mathbf{j}\right) + \frac{1}{\omega}\sin\omega t(\omega B\mathbf{j})$$

$$+ \frac{g}{\omega^2}(1 - \cos\omega t)\mathbf{i}$$

$$= \frac{g}{\omega^2}\mathbf{i} + (A\cos\omega t + B\sin\omega t)\mathbf{j},$$

so the particle oscillates simple-harmonically along a line parallel to the horizontal $y$-axis, passing through the point $\dfrac{g}{\omega^2}\mathbf{i}$.

**(b)** If $\mathbf{r}(0) = \dfrac{g}{\omega^2}\mathbf{i} + A\mathbf{j}$ and $\dot{\mathbf{r}}(0) = \omega A\mathbf{k}$

$$\mathbf{r}(t) = \cos\omega t\left(\frac{g}{\omega^2}\mathbf{i} + A\mathbf{j}\right) + \frac{1}{\omega}\sin\omega t(\omega A\mathbf{k})$$

$$+ \frac{g}{\omega^2}(1 - \cos\omega t)\mathbf{i}$$

$$= \frac{g}{\omega^2}\mathbf{i} + A\cos\omega t\,\mathbf{j} + A\sin\omega t\,\mathbf{k}.$$

Comparing this with Exercise 1, it follows that the particle moves with constant speed $\omega|A|$ in a circle of radius $|A|$ that lies parallel to the horizontal $y,z$-plane and is centred at $\dfrac{g}{\omega^2}\mathbf{i}$.

**(c)** If $\mathbf{r}(0) = \dfrac{g}{\omega^2}\mathbf{i} + A\mathbf{j}$ and $\dot{\mathbf{r}}(0) = \omega A\mathbf{i}$,

$$\mathbf{r}(t) = \cos\omega t\left(\frac{g}{\omega^2}\mathbf{i} + A\mathbf{j}\right) + \frac{1}{\omega}\sin\omega t(\omega A\mathbf{i})$$

$$+ \frac{g}{\omega^2}(1 - \cos\omega t)\mathbf{i}$$

$$= \left(\frac{g}{\omega^2} + A\sin\omega t\right)\mathbf{i} + A\cos\omega t\,\mathbf{j},$$

so the particle moves with constant speed $\omega|A|$ in a circle of radius $|A|$ that lies in the vertical $x,y$-plane and is centred at $\dfrac{g}{\omega^2}\mathbf{i}$.

**3.** The total force acting on the particle is now
$$\mathbf{F} = \mathbf{F}_B + \mathbf{F}_g$$
$$= qb(\dot{y}\mathbf{i} - \dot{x}\mathbf{j}) + mg(-\mathbf{j})$$
so Newton's second law gives
$$\ddot{\mathbf{r}} = \frac{1}{m}\mathbf{F} = \Omega(\dot{y}\mathbf{i} - \dot{x}\mathbf{j}) - g\mathbf{j}$$
where $\Omega = \dfrac{qb}{m} = 1000$, as before.

Taking each component separately, I have
$$\ddot{x} = \Omega\dot{y};$$
$$\ddot{y} = -\Omega\dot{x} - g;$$
$$\ddot{z} = 0.$$

These are identical to Equations (12a)–(12c) in the text, except for the additional term, $-g$, which describes the effects of gravity. The method for unravelling the equations is the same as before:

$$\ddot{x} = \Omega \ddot{y}$$
$$= \Omega(-\Omega \dot{x} - g)$$

so
$$\frac{d^2}{dt^2}(\dot{x}) + \Omega^2 \dot{x} = -\Omega g.$$

This is a second-order differential equation for the $x$-component of the velocity. Its general solution is:

$$\dot{x}(t) = C\cos\Omega t + D\sin\Omega t - \frac{g}{\Omega}$$

where $C$ and $D$ are arbitrary constants.

The $y$-component of the velocity is then found from

$$\dot{y}(t) = \frac{1}{\Omega}\ddot{x}(t)$$

$$= \frac{1}{\Omega}\frac{d}{dt}\left(C\cos\Omega t + D\sin\Omega t - \frac{g}{\Omega}\right)$$

$$= -C\sin\Omega t + D\cos\Omega t$$

and the $z$-component of the velocity is:

$$\dot{z}(t) = H$$

where $H$ is an arbitrary constant.

Thus, the velocity of the particle is:

$$\dot{\mathbf{r}}(t) = \left(C\cos\Omega t + D\sin\Omega t - \frac{g}{\Omega}\right)\mathbf{i}$$
$$+ (-C\sin\Omega t + D\cos\Omega t)\mathbf{j} + H\mathbf{k}.$$

Comparing this with the given initial condition, $\dot{\mathbf{r}}(0) = 2000\mathbf{i}$, I conclude that $H = 0$, $D = 0$ and

$$C = 2000 + \frac{g}{\Omega}.$$

But $\dfrac{g}{\Omega} = \dfrac{9.8}{1000} \simeq 0.01$

so $C \simeq 2000 + 0.01 \simeq 2000$

and $\dot{\mathbf{r}}(t) \simeq (2000\cos(1000t) - 0.01)\mathbf{i} - 2000\sin(1000t)\mathbf{j}.$   (S1)

Integrating both sides of this equation gives

$$\mathbf{r}(t) \simeq (2\sin(1000t) - 0.01t)\mathbf{i} + 2\cos(1000t)\mathbf{j} + \mathbf{G}$$

where $\mathbf{G}$ is a constant vector.

Comparing this with the given initial condition $\mathbf{r}(0) = 2\mathbf{j}$, I conclude that $\mathbf{G} \simeq \mathbf{0}$ so that

$$\mathbf{r}(t) \simeq (2\sin(1000t) - 0.01t)\mathbf{i} + 2\cos(1000t)\mathbf{j}.$$   (S2)

Equations (S1) and (S2) contain a real surprise. If there were no gravitational force the particle would undergo uniform circular motion in the vertical $x,y$-plane. Now the gravitational force acts down the $y$-axis, so one might expect the particle to move round a circle whose centre accelerates down the $y$-axis. But we have just shown that this is not the case: according to Equations (S1) and (S2), the centre of the circle drifts along the horizontal $x$-axis.

And that is exactly what is seen experimentally; a striking success for the tenets of this unit — the law of addition of forces and Newton's second law.

## Solutions to the exercises in Section 3

**1.(i)** $B$ and $C$: only the particles that experience zero total force could be static (i.e. *permanently* at rest).

**(ii)** None of the particles *must* be static. Particles $B$ and $C$ experience zero total force and so have no acceleration, but they could move at constant speed in a straight line.

**(iii)** All the particles could be *momentarily* at rest; whether they are or not depends on the initial conditions.

**(iv)** None of the particles *must* be momentarily at rest — again, this depends on the initial conditions.

**2.**

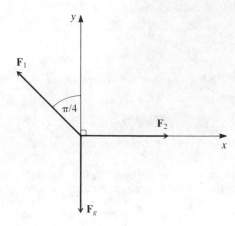

**(i)** From the above diagram,

$$\hat{\mathbf{F}}_1 = \cos\left(\frac{\pi}{4} + \frac{\pi}{2}\right)\mathbf{i} + \cos\left(\frac{\pi}{4}\right)\mathbf{j}$$

$$= -\sin\left(\frac{\pi}{4}\right)\mathbf{i} + \cos\left(\frac{\pi}{4}\right)\mathbf{j}$$

$$= -\frac{1}{\sqrt{2}}\mathbf{i} + \frac{1}{\sqrt{2}}\mathbf{j}.$$

(This could also be obtained directly, using trigonometry.)

$$\hat{\mathbf{F}}_2 = \mathbf{i}$$

and
$$\hat{\mathbf{F}}_g = -\mathbf{j}.$$

The magnitudes of $\mathbf{F}_1$ and $\mathbf{F}_2$ are, as yet, unknown, but

$$|\mathbf{F}_g| = mg.$$

Combining these results I obtain

$$\mathbf{F} = \mathbf{F}_1 + \mathbf{F}_2 + \mathbf{F}_g$$

$$= |\mathbf{F}_1|\hat{\mathbf{F}}_1 + |\mathbf{F}_2|\hat{\mathbf{F}}_2 + |\mathbf{F}_g|\hat{\mathbf{F}}_g$$

$$= |\mathbf{F}_1|\left(-\frac{1}{\sqrt{2}}\mathbf{i} + \frac{1}{\sqrt{2}}\mathbf{j}\right) + |\mathbf{F}_2|\mathbf{i} + mg(-\mathbf{j})$$

$$= \left(|\mathbf{F}_2| - \frac{1}{\sqrt{2}}|\mathbf{F}_1|\right)\mathbf{i} + \left(\frac{1}{\sqrt{2}}|\mathbf{F}_1| - mg\right)\mathbf{j}.$$

**(ii)** When the particle is static, $\mathbf{F} = \mathbf{0}$, so

$$|\mathbf{F}_2| - \frac{1}{\sqrt{2}}|\mathbf{F}_1| = 0$$

and
$$\frac{1}{\sqrt{2}}|\mathbf{F}_1| - mg = 0.$$

Thus    $|\mathbf{F}_1| = mg\sqrt{2}$

and    $|\mathbf{F}_2| = mg.$

**3.(i)** According to Equation (2) in Example 2, the total force acting on the glider is:

$$\mathbf{F} = mg\sin\phi\,\mathbf{i} + (|\mathbf{F}_N| - mg\cos\phi)\mathbf{j}$$

where $|\mathbf{F}_N| = mg\cos\phi$. Thus $\mathbf{F} = mg\sin\phi\,\mathbf{i}$.

**(ii)** By Newton's second law,

$$m\ddot{\mathbf{r}} = \mathbf{F} = mg\sin\phi\,\mathbf{i}$$

so    $\ddot{\mathbf{r}} = g\sin\phi\,\mathbf{i}$

$$= 9.8\sin(0.0061)\mathbf{i}$$

$$\simeq 0.06\,\mathbf{i}.$$

That is, the acceleration vector of the glider has magnitude $0.06\,\mathrm{m\,s^{-2}}$ and points along the $x$-axis, directly down the slope of the table.

**4.** According to Equation (3) in Example 3, the total force on the brick is

$$\mathbf{F} = (mg\sin\phi - |\mathbf{F}_f|)\mathbf{i} + (|\mathbf{F}_N| - mg\cos\phi)\mathbf{j}.$$

Setting $\phi = \dfrac{\pi}{4}$, this becomes

$$\mathbf{F} = \left(\frac{mg}{\sqrt{2}} - |\mathbf{F}_f|\right)\mathbf{i} + \left(|\mathbf{F}_N| - \frac{mg}{\sqrt{2}}\right)\mathbf{j}.$$

Since the brick is supported by the table, the $y$-component of the total force must be zero and

$$|\mathbf{F}_N| = \frac{mg}{\sqrt{2}}.$$

Since the brick is moving and the coefficient of kinetic friction is $\mu' = 0.9$,

$$|\mathbf{F}_f| = \mu'|\mathbf{F}_N| = 0.9 \times \frac{mg}{\sqrt{2}}.$$

Thus, the total force on the brick is:

$$\mathbf{F} = \left(\frac{mg}{\sqrt{2}} - 0.9\frac{mg}{\sqrt{2}}\right)\mathbf{i} = 0.1\frac{mg}{\sqrt{2}}\mathbf{i}.$$

Newton's second law then gives:

$$|\ddot{\mathbf{r}}| = \frac{1}{m}|\mathbf{F}| = 0.1 \times \frac{9.8}{\sqrt{2}} \simeq 0.69.$$

So the magnitude of the brick's acceleration is $0.69\,\mathrm{ms^{-2}}$.

**5.**

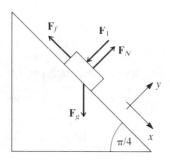

The forces acting on the brick can be quantified as in Example 3. I find that:

$$\mathbf{F}_g = mg\left(\sin\frac{\pi}{4}\mathbf{i} - \cos\frac{\pi}{4}\mathbf{j}\right)$$

$$= mg\left(\frac{1}{\sqrt{2}}\mathbf{i} - \frac{1}{\sqrt{2}}\mathbf{j}\right),$$

$$\mathbf{F}_N = |\mathbf{F}_N|\mathbf{j},$$

$$\mathbf{F}_f = |\mathbf{F}_f|(-\mathbf{i}),$$

and I also find

$$\mathbf{F}_1 = |\mathbf{F}_1|(-\mathbf{j}).$$

Thus, the total force acting on the brick is:

$$\mathbf{F} = \mathbf{F}_g + \mathbf{F}_N + \mathbf{F}_f + \mathbf{F}_1$$

$$= \left(\frac{1}{\sqrt{2}}mg - |\mathbf{F}_f|\right)\mathbf{i} + \left(|\mathbf{F}_N| - |\mathbf{F}_1| - \frac{1}{\sqrt{2}}mg\right)\mathbf{j}.$$

In order for the brick to remain static, $\mathbf{F}$ must be equal to the zero vector, so

$$|\mathbf{F}_f| = \frac{1}{\sqrt{2}}mg$$

and $|\mathbf{F}_N| = |\mathbf{F}_1| + \dfrac{1}{\sqrt{2}}mg.$

But $|\mathbf{F}_f|$ can only reach a maximum value of $\mu|\mathbf{F}_N|$, so we must have:

$$|\mathbf{F}_f| \leqslant \mu|\mathbf{F}_N|$$

i.e. $\dfrac{1}{\sqrt{2}}mg \leqslant \mu\left(|\mathbf{F}_1| + \dfrac{1}{\sqrt{2}}mg\right)$

which gives

$$|\mathbf{F}_1| \geqslant \frac{1}{\sqrt{2}}mg\left(\frac{1}{\mu} - 1\right).$$

Putting $m = 0.8$, $g = 9.8$ and $\mu = 0.5$, I obtain

$$|\mathbf{F}_1| \geqslant \frac{1}{\sqrt{2}}mg \simeq 5.54$$

So the *minimum* force that must be applied in a direction perpendicular to the table is 5.54 newtons. (Notice that the brick is held in place because $\mathbf{F}_1$ increases the normal reaction force, and this, in turn, allows the frictional force to increase.)

## Solutions to the exercises in Section 4

**1.** Using Equation (6) with $g$ replaced by 0.06 I find that:

$$\frac{0.06}{2v_0^2\cos^2\theta} = 1$$

and

$$\tan\theta = 1.6$$

Thus $\qquad \theta \simeq 1.01$ radians $\quad (\simeq 58$ degrees$)$

and $\qquad v_0 \simeq \sqrt{\dfrac{0.06}{2\cos^2(1.01)}} \simeq 0.33.$

So the initial speed of the puck was $0.33\,\mathrm{ms^{-1}}$ and the initial angle between its direction of motion and the $x$-axis was about 58 degrees.

**2.(i)** $v_{max} = \sqrt{gR_{max}} \simeq \sqrt{9.8 \times 22.15} \simeq 14.73.$

**(ii)** We must calculate the maximum height of the shot. Since the optimum angle is 45° and the maximum range is $R_{max} = \dfrac{v_{max}^2}{g}$, Equation (6) can be simplified to:

$$y = -\frac{x^2}{R_{max}} + x.$$

The *maximum* height occurs when $\dfrac{dy}{dx} = 0$

i.e. $\qquad 0 = \dfrac{-2x}{R_{max}} + 1$

and $x = \dfrac{R_{max}}{2}$, as suggested (not proved!) by Figure 2.

Inserting this value of $x$ into our equation for $y$ gives:

$$y = -\frac{(R_{max}/2)^2}{R_{max}} + \frac{R_{max}}{2} = \frac{R_{max}}{4}.$$

Thus, the maximum height of the shot is $\dfrac{22.15}{4} \simeq 5.54$ metres.

The ceiling of the gymnasium is too low.

**3.(i)** From Equation (13)

$$\tan\theta = \frac{L}{R_{max}}.$$

Substituting this expression into Equation (12), I find

$$0 = -\frac{R_{max}^2}{2L}\left(1 + \frac{L^2}{R_{max}^2}\right) + L + h$$

so $\dfrac{R_{max}^2}{2L} = -\dfrac{L}{2} + L + h$

and $R_{max} = \sqrt{L^2 + 2hL} = L\sqrt{1 + \dfrac{2h}{L}}.$

**(ii)** Substituting this expression for $R_{max}$ into Equation (13) gives:

$$\tan\theta = \frac{L}{R_{max}} = \frac{1}{\sqrt{1 + \dfrac{2h}{L}}}.$$

**4.** From the diagram below, the direction of release bisects the angle $CAB$ if:

$$\theta + \phi = \frac{1}{2}\left(\frac{\pi}{2} + \phi\right)$$

i.e. if $2\theta = \dfrac{\pi}{2} - \phi$

so that $\tan 2\theta = \cot\phi = \dfrac{R_{max}}{h}.$

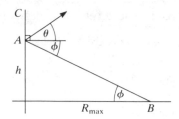

However,

$$\tan 2\theta = \frac{2\tan\theta}{1 - \tan^2\theta}$$

So Equation (15) gives:

$$\tan 2\theta = \frac{\dfrac{2}{\sqrt{1 + \dfrac{2h}{L}}}}{1 - \dfrac{1}{1 + \dfrac{2h}{L}}} = \frac{2\sqrt{1 + \dfrac{2h}{L}}}{1 + \dfrac{2h}{L} - 1}$$

$$= \frac{L\sqrt{1 + \dfrac{2h}{L}}}{h} = \frac{R_{max}}{h}, \quad \text{as required.}$$

**5.(i)** Some care is needed here. Presumably, the world champion launches at the *optimum* angle rather than 45°, so the record of 22.15 metres corresponds to $R_{max}$ rather than to $L$. We can find $L$ with the aid of Equation (14):

$$R_{max} = L\sqrt{1 + \frac{2h}{L}}$$

so $L^2 + 2hL - R_{max}^2 = 0$

and $L = \dfrac{-2h + \sqrt{4h^2 + 4R_{max}^2}}{2}$

$= \sqrt{R_{max}^2 + h^2} - h.$

In the case of the world champion,

$$L = \sqrt{(22.15)^2 + 2^2} - 2 \simeq 20.24.$$

Inserting this value of $L$ into Equation (15), we find:

$$\tan\theta \simeq \frac{1}{\sqrt{1 + \dfrac{2 \times 2}{20.24}}} \simeq 0.914$$

This gives an optimum angle of $\theta \simeq 42.4°$.

(An alternative method of solution is provided by the result of Exercise 4. The optimum angle satisfies

$$\tan 2\theta = \frac{R_{max}}{h} = \frac{22.15}{2}$$

so $\theta = \frac{1}{2}\arctan\dfrac{22.15}{2} \simeq 42.4°$).

**(ii)** I *did* launch from 45°, but not from zero height.

Using Equation (12),

$$0 = \frac{-R^2}{L} + R + h$$

so $L = \dfrac{R^2}{R + h} = \dfrac{25}{6.5}.$

Thus, my *maximum* range is:

$$R_{max} = L\sqrt{1 + \frac{2h}{L}}$$

$$= \frac{25}{6.5}\sqrt{1 + \frac{2 \times 1.5 \times 6.5}{25}}$$

$$\simeq 5.13$$

and my optimum angle of launch is given by

$$\tan\theta = \frac{L}{R_{max}} = \frac{25}{6.5} \times \frac{1}{5.13} \simeq 0.75$$

so that my maximum range is 5.13 metres and my optimum angle of release is 36.9°.

## Solutions to the exercises in Section 5

**1(i)** If $x = x_0$, a constant, Equation (3a) gives

$$0 = mg - \frac{T}{l}x_0$$

so $T = \dfrac{mgl}{x_0}$

and the tension in the string is constant.

**(ii)** Equations (3b) and (3c) then take the form

$$\ddot{y} + \omega^2 y = 0$$

and $\quad \ddot{z} + \omega^2 z = 0$

where $\quad \omega^2 = \dfrac{T}{ml} = \dfrac{g}{x_0}.$

These are the equations of simple harmonic motion and their general solutions are:

$$y(t) = C\cos\omega t + D\sin\omega t,$$

$$z(t) = G\cos\omega t + H\sin\omega t,$$

where $C$, $D$, $G$ and $H$ are arbitrary constants. The time taken to complete one orbit is the time that elapses before $y$ and $z$ return to their initial values. Since

$$y\left(t + \frac{2\pi}{\omega}\right) = y(t)$$

and

$$z\left(t + \frac{2\pi}{\omega}\right) = z(t),$$

this time is:

$$\frac{2\pi}{\omega} = 2\pi \sqrt{\frac{x_0}{g}}.$$

**2.(i)** From Equation (10)

$$E = \tfrac{1}{2}ml^2\dot\theta^2 + mgl\,(1 - \cos\theta).$$

Thus Equation (12) gives:

$$T = mg\cos\theta + ml\dot\theta^2$$

$$= mg\cos\theta + \frac{2}{l}(\tfrac{1}{2}ml^2\dot\theta^2)$$

$$= mg\cos\theta + \frac{2}{l}(E - mgl(1 - \cos\theta))$$

$$= mg(3\cos\theta - 2) + \frac{2E}{l}.$$

**(ii)** The bob is released from rest at $\theta = \dfrac{\pi}{2}$, so

$$E = 0 + mgl\left(1 - \cos\frac{\pi}{2}\right) = mgl$$

and $T = mg\,(3\cos\theta - 2) + 2mg = 3mg\cos\theta$.

The maximum value of $T$ is $3mg$ (this occurs when $\theta = 0$, at the bottom of the swing). The string must be strong enough to exert this tension.

**(iii)** The string breaks when $T = 1.5\,mg$. This occurs at an angle $\theta$ given by:

$$1.5mg = 3mg\cos\theta$$

so $\cos\theta = 0.5$ and $\theta = \dfrac{\pi}{3}$ radians (i.e. 60°) to the vertical.

**3.** The total energy of the bob is fixed by the initial conditions:

$$E = \tfrac{1}{2}mv_0^2 + mgl(1 - \cos 0) = \tfrac{1}{2}mv_0^2$$

and the tension in the string is:

$$T = mg\,(3\cos\theta - 2) + \frac{2E}{l}$$

$$= mg\,(3\cos\theta - 2) + \frac{mv_0^2}{l}.$$

**(i)** If the string is never to become slack, $T > 0$ for all values of $\theta$, so

$$mg\,(3\cos\theta - 2) + \frac{mv_0^2}{l} > 0$$

and $\quad v_0^2 > gl(2 - 3\cos\theta) \quad$ for all $\theta$.

The largest value of $gl(2 - 3\cos\theta)$ is $5gl$ (this occurs when $\theta = \pi$, at the top of the swing). So, in order for the bob to loop the loop without the string going slack we must have

$$v_0^2 > 5gl$$

i.e. $\qquad v_0 > \sqrt{5gl}$

(e.g. for a one metre pendulum, $v_0 > 7\,\text{ms}^{-1}$ approximately).

**(ii)** If $v_0 = 2\sqrt{gl}$,

$$T = mg\,(3\cos\theta - 2) + \frac{4mgl}{l}$$

$$= mg\,(3\cos\theta + 2).$$

The string goes slack when $T = 0$, i.e. when $\cos\theta = -\tfrac{2}{3}$

and $\qquad \theta \simeq 2.3$ radians $\simeq 132°$.

**4.** We have established that:

$$E = \tfrac{1}{2}m|\dot{\mathbf{r}}|^2 + mg(-x) = \text{constant}.$$

For the particle in Example 1, $x = x_0$ is constant so

$$|\dot{\mathbf{r}}| = \sqrt{\frac{2}{m}(E + mg\,x_0)}$$ is also constant and the circular motion is *uniform*.

**5.(i)** Equation (16) applies in this case because the only force to consider, apart from gravity, is the normal reaction force, $\mathbf{F}_N$. But $\mathbf{F}_N$ acts perpendicular to the wire, while the particle's velocity $\mathbf{v}$ is tangential to the wire, so $\mathbf{F}_N \cdot \mathbf{v} = 0$.

**(ii)** Equation (16) does not apply in this case because the forces of friction and air resistance are *not* perpendicular to $\mathbf{v}$.

**(iii)** Equation (16) applies in this case because $\mathbf{F}_B = q\,(\mathbf{v} \times \mathbf{B})$ is perpendicular to $\mathbf{v}$.

# Appendix 2: Solutions to the problems

## Solutions to the problems in Section 6

**1.**

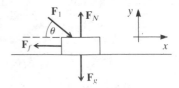

My diagram is shown above. I have chosen my $y$-axis to point vertically upwards and my $x$-axis so that all the forces act in the $x,y$-plane. In this co-ordinate system the forces acting on the brick are

$$\mathbf{F}_g = mg(-\mathbf{j});$$

$$\mathbf{F}_N = |\mathbf{F}_N|\mathbf{j};$$

$$\mathbf{F}_f = |\mathbf{F}_f|(-\mathbf{i});$$

$$\mathbf{F}_1 = |\mathbf{F}_1|(\cos\theta\,\mathbf{i} - \sin\theta\,\mathbf{j}).$$

The total force acting on the brick is therefore

$$\mathbf{F} = \mathbf{F}_g + \mathbf{F}_N + \mathbf{F}_f + \mathbf{F}_1$$

$$= (|\mathbf{F}_1|\cos\theta - |\mathbf{F}_f|)\mathbf{i} + (|\mathbf{F}_N| - |\mathbf{F}_1|\sin\theta - mg)\mathbf{j}.$$

For small enough values of $|\mathbf{F}_1|$ the brick is static, $\mathbf{F} = \mathbf{0}$, so

$$|\mathbf{F}_f| = |\mathbf{F}_1|\cos\theta;$$

$$|\mathbf{F}_N| = |\mathbf{F}_1|\sin\theta + mg.$$

Since $|\mathbf{F}_f| \leqslant \mu|\mathbf{F}_N|$, the brick remains static if

$$|\mathbf{F}_1|\cos\theta \leqslant \mu(|\mathbf{F}_1|\sin\theta + mg).$$

It can only move if

$$|\mathbf{F}_1|\cos\theta > \mu(|\mathbf{F}_1|\sin\theta + mg).$$

In order to move the brick we must certainly have

$$|\mathbf{F}_1|(\cos\theta - \mu\sin\theta) > \mu mg > 0.$$

But this is not possible if $\cos\theta - \mu\sin\theta < 0$, so the brick will

not budge if $\mu > \dfrac{\cos\theta}{\sin\theta} = \cot\theta$.

**2.**

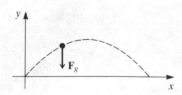

My diagram is shown above. I have chosen my $y$-axis to point vertically upwards and my $x$-axis so that the arrow moves in the $x,y$-plane. Then, ignoring air resistance, Newton's second law gives

$$\ddot{x} = 0;$$

$$\ddot{y} = -g;$$

$$\ddot{z} = 0.$$

The first of these equations can be integrated to give

$$\dot{x}(t) = \text{constant}.$$

The value of the constant depends on the way the arrow is launched. Let the launch speed be $v_0$ and the launch angle be $\pi/4$ radians (for maximum range). Then

$$\dot{x}(0) = v_0\cos\left(\frac{\pi}{4}\right) = \frac{v_0}{\sqrt{2}}$$

so

$$\dot{x}(t) = \frac{v_0}{\sqrt{2}}$$

and Hiawatha must run at least as fast as this if he is to catch up or overtake the arrow.

To estimate the speed, $v_0$, of an arrow from Hiawatha's bow, I use the information that an arrow, fired vertically, remains in the air for at least 9 seconds.

Integrating the equation for $\ddot{y}$, I obtain

$$\dot{y}(t) = -gt + B$$

and $\qquad y(t) = -\tfrac{1}{2}gt^2 + Bt + C.$

For simplicity, I neglect Hiawatha's height and use the initial conditions, $y(0) = 0$, $\dot{y}(0) = v_0$.

These give $C = 0$ and $B = v_0$ so that

$$y(t) = -\tfrac{1}{2}gt^2 + v_0t$$

and the time, $T$, spent by the arrow in the air is given by

$$0 = -\tfrac{1}{2}gT^2 + v_0T.$$

Hence $T = \dfrac{2v_0}{g} = 9$

so $v_0 = \dfrac{9.8 \times 9}{2} = 44.1$

and Hiawatha can run at $\dfrac{44.1}{\sqrt{2}} \simeq 31.2\,\text{m s}^{-1}.$

Swift of foot indeed!

**3.** Let $m$ be the mass of the car,

$d$ be its distance from the wall when evasive action is started,

$v_0$ be the speed of the car at this moment,

and suppose the motorist applies the maximum possible braking force, of constant magnitude $|\mathbf{F}|$. Consider, first, the strategy of steering the car with constant speed $v_0$ in the arc of a circle of radius $l$. According to Example 1 and Exercise 1 of Section 2 we have

$$|\mathbf{F}| = ml\omega^2$$

where $\qquad l\omega = v_0$

so $\qquad |\mathbf{F}| = \dfrac{mv_0^2}{l}$

and the car will avoid the wall if

$$l = \frac{mv_0^2}{|\mathbf{F}|} < d.$$

On the other hand, if the motorist travels in a straight line directly towards the wall, we can use the methods of *Unit 4*. Suppose the car travels along the $x$-axis in the sense of increasing $x$. Let the brakes be applied at $t = 0$ and $x = 0$ and let the car stop at $t = T$ and $x = X$. Then

$$\ddot{x}(t) = -\frac{|\mathbf{F}|}{m}$$

so $\qquad \dot{x}(t) = -\dfrac{|\mathbf{F}|}{m}t + v_0$

and $\qquad x(t) = -\dfrac{1}{2}\dfrac{|\mathbf{F}|}{m}t^2 + v_0t.$

When the car comes to rest we have

$$0 = -\frac{|\mathbf{F}|}{m}T + v_0$$

and $\qquad X = -\dfrac{1}{2}\dfrac{|\mathbf{F}|}{m}T^2 + v_0T$

$$= -\frac{1}{2}\frac{|\mathbf{F}|}{m}\left(\frac{mv_0}{|\mathbf{F}|}\right)^2 + \frac{mv_0^2}{|\mathbf{F}|}$$

$$= \frac{1}{2}\frac{mv_0^2}{|\mathbf{F}|}.$$

The car will therefore avoid the wall if

$$X = \frac{1}{2}\frac{mv_0^2}{|\mathbf{F}|} < d.$$

Note that $X = \tfrac{1}{2}l$, so it would be more sensible for the motorist to steer straight at the wall than to attempt to swing the car round at constant speed.

**4.(i)** The initial kinetic energy of the bob is

$$\tfrac{1}{2}m(2\sqrt{gl})^2 = 2mgl.$$

When the bob is stationary at the top of its flight, this has all been converted into potential energy, mass × g × height. The highest point attained by the bob is therefore a distance $2l$ above its starting point. This agrees with Figure 3.

**(ii)** Using the result of Exercise 3(ii) in Section 5, the string becomes slack when $\cos\theta = -\tfrac{2}{3}$. The height of the particle at this point is

$$l(1 - \cos\theta) = \frac{5}{3}l.$$

The rest of the motion, up to the highest point, must be calculated as for a shot. But first, we need to know the bob's speed and direction of motion, just as the string becomes slack.

The speed can be found from the conservation of energy:

$$E = \tfrac{1}{2}m|\mathbf{v}|^2 + mgl(1 - \cos\theta)$$

so $\qquad 2mgl = \tfrac{1}{2}m|\mathbf{v}|^2 + mgl\left(\dfrac{5}{3}\right)$

and $\qquad |\mathbf{v}| = \sqrt{\dfrac{2}{3}gl}.$

The angle can be found by simple geometry (see the figure below). It is $(\pi - \theta)$.

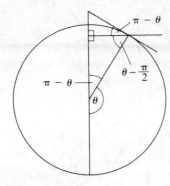

To find the highest point reached by the particle, I shall set up a $y$-axis which points vertically upwards and whose origin is at the point where the string goes slack. Measuring time from this moment also, we have

$$y = -\tfrac{1}{2}gt^2 + (u\sin\phi)t$$

where $u = \sqrt{\dfrac{2}{3}gl}$ and $\sin\phi = \sin(\pi - \theta) = \sin\theta$.

The highest point occurs when $\dfrac{dy}{dt} = 0$, that is when

$$0 = -gt + u\sin\phi$$

so $$t = \frac{u\sin\phi}{g}$$

and $$y = -\tfrac{1}{2}g\left(\frac{u\sin\phi}{g}\right)^2 + \frac{(u\sin\phi)^2}{g} = \frac{u^2\sin^2\phi}{2g}.$$

Thus the extra height gained by the particle while the string is slack is

$$y = \frac{1}{2g} \times \frac{2}{3}gl(1 - \cos^2\theta) = \frac{l}{3} \times \frac{5}{9}.$$

The total height reached by the particle above its static height is therefore

$$\frac{5}{3}l + \frac{5}{27}l = \frac{50}{27}l.$$